ALDEHYDES IN BIOLOGICAL SYSTEMS
THEIR NATURAL OCCURRENCE AND BIOLOGICAL ACTIVITIES

E. Schauenstein H. Esterbauer
H. Zollner   Translator P.H. Gore

Series editor J.R.Lagnado

1 **Free radical mechanisms in tissue injury** T.F.Slater
2 **Biochemical reactors** B.Atkinson
3 **Brain function and macromolecular synthesis** B.Jakoubek
4 **Infectious multiple drug resistance** S.Falkow
5 **Aldehydes in biological systems** E.Schauenstein, H.Esterbauer, H.Zollner

ALDEHYDES IN BIOLOGICAL SYSTEMS
THEIR NATURAL OCCURRENCE AND BIOLOGICAL ACTIVITIES

E. Schauenstein H. Esterbauer
H. Zollner Translator P.H. Gore

Pion Limited, 207 Brondesbury Park, London NW2 5JN

© 1977 Pion Limited

All rights reserved. No part of this book may be reproduced in any form by photostat microfilm or any other means without written permission from the publishers.

ISBN 0 85086 059 8

Printed in Great Britain

Preface

This book originated in a proposal put to me by Pion Ltd to write a monograph on hydroxyalkenals, the subject of my specific research. Whilst being flattered by the invitation, I found I was of the opinion that it would be better if the subject were discussed in relation to the relevant findings of other researchers in this field, as well as to the biochemical and biological studies on aldehydes in general. The publishers fully agreed with this suggestion, and so I commenced by securing the invaluable coauthorship of Professor H Esterbauer and Dr H Zollner, after which we were able to begin work.

At a very early stage it was realised that the book would go far beyond its original scope, owing to the wide distribution of aldehydes in nature and to their long history in biochemistry and biology. From the historical point of view, it will no doubt be recalled that formaldehyde is present in interstellar space as one of the constituents of 'dark clouds' and that aldehydes and hydrogen cyanide are formed in the well-known model systems which have been set up to simulate probable prebiotic conditions. Since amino acids have also been detected in these systems—in which mixtures of methane and ammonia over boiling water are exposed to electric discharges—it may be assumed that these amino acids were formed from aldehydes and hydrogen cyanide by means of the Strecker synthesis, and this suggests that aldehydes may have played an important role in the formation of proteins during the prebiotic phase of the earth's development.

It also seems likely that the simplest aldehyde, formaldehyde, was starting material for the prebiotic formation of sugars. Support for this idea comes from the fact that, at a temperature of $100°C$, clays such as kaolin, catalyze the formation of monosaccharides, including riboses, in good yield from dilute aqueous solutions of formaldehyde with glycoaldehyde and glyceraldehyde appearing as important intermediates in the process. Thus aldehydes may well have been implicated in the prebiotic formation of the sugar component of nucleic acids.

In addition to their basic function in the formation of terrestrial life, we ought to be particularly impressed today by two characteristics of aldehydes: their wide distribution in living matter and their wide range of biochemical and biological functions, ranging from their occurrence as metabolic intermediates, to their more or less selective interference in energy metabolism and biosynthesis, to highly specific biological effects. The purpose of this book is to bring all such facts to the attention of the reader in as comprehensive a manner as possible.

Yet another point which it is important to mention is that a large number of aldehydes occur under abnormal physiological conditions— during the processing of foods and beverages, the autoxidation of unsaturated lipids, the irradiation of sugars, and in various combustion processes. Many of these aldehydes are highly reactive substances and are potent inhibitors of metabolism and of cell growth. Thus aldehydes play an important role in environmental-pollution problems too.

Preface

The account is intended to be as complete as we can reasonably make it and as such we hope it will be of use to senior undergraduates, graduates, and research workers in chemistry, biochemistry, and biology, especially those working in the field of cancer research.

This preface would be altogether incomplete if I failed to express my most sincere thanks to Professor T F Slater of Brunel University, who first suggested that I write a short monograph reviewing our research on the biological effects of hydroxyaldehydes, and to Pion Ltd for their invitation to write this book as well as for their most helpful and understanding cooperation. I am particularly grateful for and appreciative of the arrangement by which Dr P H Gore of Brunel University, endowed with expert technical knowledge and a fine sense of linguistic perception, and with great forbearance, undertook to render into English the original German version of the manuscript.

With regard to our own research on hydroxyalkenals I want to thank first of all my friend Wilhelm von Gwinner, Munich, Germany, whose financial aid enabled us to start our work in 1953 and to proceed with it until we were given more substantial grants, first from the Rockefeller Foundation, New York, and from the Department of Health, Education, and Welfare, Public Health Service, Bethesda, Maryland, and later from the Fonds zur Förderung der wissenschaftlichen Forschung, Vienna, Austria. For this support, too, I should like to express my most heartfelt gratitude. My thanks are also due to all the colleagues with whom I had the honour and pleasure to work, and last but not least to Mrs. Liane Golznig, MA, of our Institute, for carefully preparing the manuscripts and for most valuable help in intricate problems of translation.

E Schauenstein
Professor of Biochemistry, University of Graz

Contents

1	**Introduction**	
1.1	Chronology	1
1.2	Occurrence of aldehydes *in vivo*	3
1.3	Reactive sites of different aldehydes	3
1.4	Sites of attack in substrates	3
1.5	Mechanisms of action	4
1.6	What will the future bring?	7

Part 1 Aldehydes as inhibitors of metabolism, biosynthesis, and cell division

2	**Saturated aldehydes**	
2.1	Structure of saturated aldehydes in water	9
2.2	Reactions of alkanals with amino acids	9
2.3	Effects of alkanals on proteins	11
	2.3.1 Special proteins: collagen	13
	2.3.2 Special proteins: serum albumin	14
	2.3.3 Special proteins: lipase	14
2.4	Some physiological effects of alkanals	15
2.5	Effect of alkanals on cell respiration	16
	2.5.1 Formaldehyde	16
	2.5.2 Other aldehydes	16
2.6	Antitumour effects of alkanals	17
2.7	Effects of alkanals on bacteria and fungi	20
3	**α,β-Unsaturated aldehydes**	
3.1	2-Alkenals	25
	3.1.1 Structure of α,β-unsaturated aldehydes	25
	3.1.2 Reactions with biologically important groups	25
	3.1.3 Toxicity of α,β-unsaturated aldehydes	32
	3.1.4 Antitumour effects of α,β-unsaturated aldehydes	35
	3.1.5 Bactericidal and fungicidal effects of α,β-unsaturated aldehydes	38
	3.1.6 Antiviral effects of α,β-unsaturated aldehydes	40
3.2	4-Hydroxy-2-alkenals	42
	3.2.1 Occurrence in autoxidized polyene fatty acids	42
	3.2.2 Biochemical and biological effects of natural 4-hydroxy-trans-2,3-octen-1-al (HOE)	46
	3.2.3 Syntheses of 4-hydroxyenals	51
	3.2.4 Biochemical and biological effects of synthetic 4-hydroxyenals	52
	3.2.5 Biochemical and biological action of 4-hydroxypentenal (HPE)	54
	3.2.6 Differences in effects of hydroxyenals (HE) on different processes in the same cell type, or on the same processes in different cell types	65
	3.2.7 Biological consequences	75
3.3	Metabolic fate of α,β-unsaturated aldehydes	88
4	**α-Hydroxyaldehydes**	
4.1	Glyceraldehyde	103

5	**Dicarbonyl compounds**	
5.1	α-Ketoaldehydes	112
	5.1.1 Structures of glyoxal and methylglyoxal	112
	5.1.2 Reactions with amines and amino acids	113
	5.1.3 Reaction with proteins	115
	5.1.4 Reactions with nucleic acids and their building blocks	118
	5.1.5 Effects on different cell functions	120
	5.1.6 Effects of α-ketoaldehydes on tumour growth	126
	5.1.7 Action of α-ketoaldehydes on the growth of bacteria	130
	5.1.8 Action of α-ketoaldehydes on viruses	131
5.2	Malonaldehyde	133
	5.2.1 General features	133
	5.2.2 Reactions of malonaldehyde with amino acids	135
	5.2.3 Reactions of malonaldehyde with nucleic acids	136
	5.2.4 Reactions of malonaldehyde with proteins and enzymes	137
	5.2.5 Effects of malonaldehyde on various cell functions	139
5.3	Glutaraldehyde	
	5.3.1 General features	140
	5.3.2 Reactions with proteins	142
	5.3.3 Effects of glutaraldehyde on various cell functions	145
	5.3.4 Action of glutaraldehyde on viruses and bacteria	146
5.4	α,β-Unsaturated carbonyl compounds from energetically irradiated aqueous sugar solutions	147
6	**Stable derivatives of aldehydes**	
6.1	Antibacterial and fungistatic effects	158
6.2	Antiviral effects	158
6.3	Carcinostatic effects	159
Part 2	**Naturally occurring aldehydes and their biological functions**	
7	**Schematic survey of the most important aldehydes involved in intermediary metabolism**	
7.1	Aliphatic long chain aldehydes from animal tissues and from serum	163
7.2	α-Ketoaldehydes	164
	7.2.1 Glyoxal	164
	7.2.2 Methylglyoxal	164
	7.2.3 Other α-ketoaldehydes	168
8	**Aldehydes with specific biological functions**	
8.1	Succinaldehydic acid	172
8.2	Indol-3-ylacetaldehyde and 5-hydroxyindol-3-ylacetaldehyde	172
8.3	Pyridoxal–pyridoxal phosphate	174
	8.3.1 Biosynthesis of pyridoxal	177
8.4	Retinal and dehydroretinal	179
8.5	Collagenaldehyde	181

Erratum

Colour plate 1. Photographs (a) and (c) have been transposed.

8.6	Aldehydes in plants and fruit	181
	8.6.1 Aldehydes from cuticular leaf wax	181
	8.6.2 2-*trans*-Hexenal	183
	8.6.3 Aldehydes in fruit	184
8.7	Aldehydes as substances transmitting information	185
	8.7.1 Biogenesis of semiochemicals	188
	8.7.2 Recognition of pheromones	188
8.8	Carbonyl compounds as vehicles of taste and odour	189
	8.8.1 Origin of flavour-producing carbonyl compounds	189
	8.8.2 Aldehydes associated with specific flavours	193
8.9	Participation of aldehydes in bioluminescence	194

Index — 201

1

Introduction

Even researchers currently working in the field of aldehydes are likely to be startled at the extent and content of the relevant literature cited in the chapters of this book. And it is surprising to realise that substances so very widely distributed are, at the same time, involved so fundamentally in biochemical processes and biological functions.

It therefore seemed a worthwhile undertaking to review the work that has been done, in an attempt to provide the reader who has a general biological interest with a more or less comprehensive survey, and in the hope that even some specialists may here and there find new information or a novel and fertile aspect for their own researches.

In an introductory section it would seem appropriate and justified to present first of all a general picture of the highlights of biological research on aldehydes, together with the historical development, though those investigations which have actually initiated research on biological effects will be mentioned. We shall restrict ourselves to only those studies on metabolic effects *in vitro*, inhibition of tumour growth *in vivo*, and bactericidal or virucidal effects which were actually the first in the field, or which appeared at a particularly early date. The detailed discussion and citation of the work will, of course, come later in the book.

1.1 Chronology

A survey of the literature reveals that the very first experiments on the inhibitions of metabolic processes were carried out with naturally occurring aldehydes *in vitro*. In 1929 Mendel reported that glyceraldehyde almost completely inhibits the glycolysis of cancer cells, but leaves respiration, as also for normal cells, quite uninfluenced. In this study the possibility is already mooted of an antitumour effect of glyceraldehyde, and this idea has occupied numerous authors from Riley and Pettigrew in 1944 right up to the present time. Experiments with the effects of aldehydes on bacteria were first described by Cutinelli, and on viruses by Mücke, as late as 1967.

Around the same time two pioneer studies on α-ketoaldehydes were made: in 1931 by Schmidt on the deactivation of diphtheria toxin by glyoxal, and in 1932 by Kisch on the inhibition of respiration of animal tissues and tumour cells by methylglyoxal. Both investigations heralded a particularly fertile period of development. The work of Schmidt was followed by a series of investigations by numerous authors on the bacteriostatic effects of α-ketoaldehydes. The work of Kisch, initially performed for no specific reason, nevertheless marks the beginning of very many investigations which have led to the following results:
(1) the recognition that α-ketoaldehydes are potent inhibitors of the proliferation of bacterial, plant, human, and animal cells;

(2) the fruitful working concept of Szent-Györgyi and Együd, that biogenic α-ketoaldehydes function as naturally occurring growth inhibitors, according to a generally valid principle (1965/1966);
(3) the recognition that α-ketoaldehydes possess antiviral properties, proceeding from the systematic investigations of Underwood (1956);
(4) the recognition of the virucidal action of α-ketoaldehydes in its turn provided the stimulus for the investigations of antitumour effects (French and Friedländer, 1958), and these have subsequently found extraordinary extensions.

Also around 1930, a period of great importance for biological research on aldehydes, Heymans reported (in 1926) on the deactivation of plant toxins with formaldehyde, this work being followed by many investigations on the deactivation of other toxins, and Strong (in 1932) began a series of investigations on the tumour-inhibiting effect of ethereal oils with heptanal as the active component. This then led to other important investigations, especially those of Boyland (1940) and Dittmar (1940), as well as those of many later authors.

It is rather remarkable that systematic investigations of inhibition of metabolism by alkanals did not follow until much later (Rehak and Truitt, 1958, and subsequent workers).

2-Alkenals as a group received attention for the first time in 1938, with the experiments of Boyland on the antitumour effects of citral, again a chief component of certain etheral oils. This theme was then taken up by many investigators for other α,β-unsaturated aldehydes. In the same year also, Ingersoll reported on bactericidal effects, a subject which was to be studied extensively in the years to follow.

As with saturated aldehydes, the metabolic effects of α,β-unsaturated aldehydes were not investigated until long after papers on their inhibitory effects on tumours, bacteria, and viruses (Nyman, 1954) had been published. As late as 1957 Stack concerned himself with the mechanism of the toxicity of such aldehydes towards microorganisms. He found that inhibition of biological oxidation occurred most strongly with acrolein. This aldehyde is a highly irritant compound which in later years became particularly topical on account of its manifold environmental pollution effects, but which excited interest in tumour research when it was shown by Alarcon (in 1971) that *cyclo*phosphamide, an agent much used in tumour therapy, develops its activity via acrolein formed as an intermediate.

4-Hydroxy-2-alkenals have not long been studied. The first representative of this group, 4-hydroxy-2-*trans*-octenal, was isolated for the first time by Schauenstein and Esterbauer in 1962/1963 from linoleic acid alkylesters, and was subsequently investigated in detail. Inhibitory effects of the compounds on energy metabolism were first reported in 1964.

1.2 Occurrence of aldehydes *in vivo*
Aldehydes which are active as inhibitors of energy metabolism, of biosyntheses and of cell division, may be found as intermediates in the metabolism of man and animals. They occur in the blood and in different organs, and also in fruit, seeds, plants, and microorganisms; they are formed by high-energy irradiation of carbohydrates or protein and by autoxidation of unsaturated fats and fatty acids. It would therefore appear justified to regard these compounds as ubiquitous.

1.3 Reactive sites of different aldehydes
When an examination is made of how the separate classes of aldehydes develop their effects, it becomes obvious that saturated aldehydes are active as inhibitors only through the aldehyde group, but that the inhibitory effects may be modified by more general properties that are a function of the structure of the molecule, such as its hydrophilic, lipophilic, or polar properties.

For aldehydes possessing a carbon–carbon double bond or a carbonyl group in the neighbourhood of the aldehyde group, the following possibilities exist in principle:
(1) The special effects may arise from the aldehyde function or from the vicinal olefinic group, as in 2-alkenals. A special case is represented by malonaldehyde, which through tautomerism converts into an enol group vicinal to the aldehyde group, and the former then represents the primary reactive grouping.
(2) The special effects are primarily caused only by a neighbouring olefinic grouping. 4-Hydroxy-2-alkenals may serve as example: here the olefinic group is activated on the one hand through the aldehyde group, and on the other hand by the hydroxyl group situated at C-4.
(3) The special effects appear to be based on the primary reaction of the aldehyde group, activated particularly by a neighbouring keto group, as in α-ketoaldehydes.

1.4 Sites of attack in substrates
The metabolic processes in which aldehydes participate, or the means by which they modify biological structures, involve on the one hand principally the sulphydryl group of nonprotein thiols as well as of protein molecules, and on the other the amino groups of amines of lower molecular weight, or of amino acids, proteins or nucleic acids.

The effects involve in general energy metabolism, the biosyntheses of nucleic acids and proteins, and mitosis. As a result they profoundly influence cell division and in particular the growth of malignant tumours, of bacteria, and of fungi, as well as the reproduction of viruses; in consequence the inhibitory effects by far outweigh other effects.

Having shown which groups in the different classes of aldehydes generally undergo reaction with which groups in the cell, in the tissues or body fluids (thereby causing the biochemical and biological effects referred to), it becomes tempting to consider the mechanisms of action of the different classes of aldehydes.

1.5 Mechanisms of action

In general terms it was found that alkanals, 2-alkenals, and α-ketoaldehydes attack sulphydryl groups as well as amino groups, but 4-hydroxy-2-alkenals, at least in the concentration range 10^{-5}–10^{-3} M and for durations of reaction of up to several hours, attack practically only sulphydryl groups.

For alkanals the known reactions comprise:

(a) Reactions with the sulphydryl groups of cysteine, formation of thiazolidinecarboxylic acids. In principle hemimercaptals as well as Schiff's bases may be formed, but it can be taken for granted that formation of the final thiazolidinecarboxylic acid takes place exclusively by cyclization of the Schiff's base formed initially (scheme 1.1).

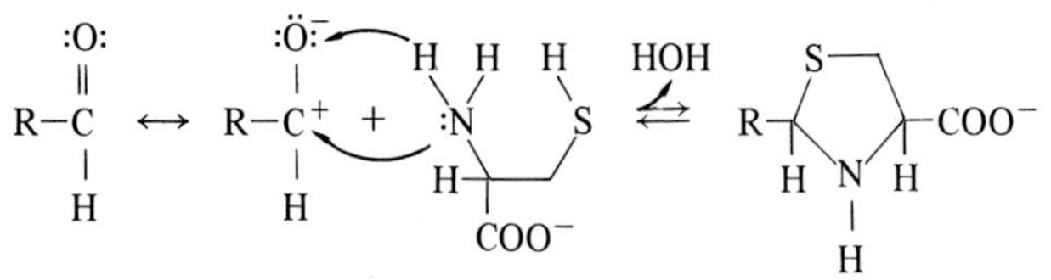

Scheme 1.1

(b) Reactions with the amino group forming Schiff's bases or imino compounds. In particular cross-linkages may form in proteins, according to scheme 1.2.

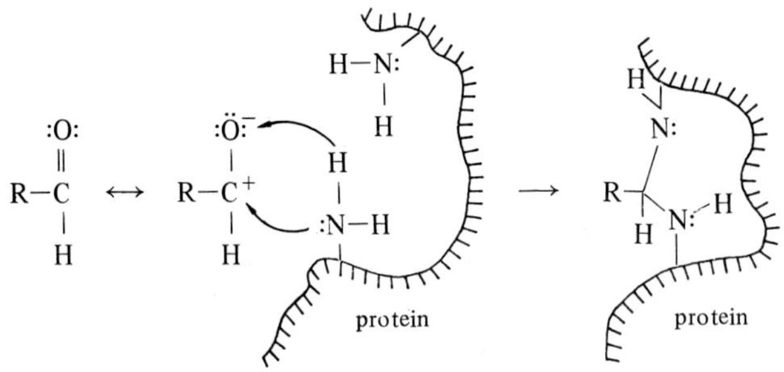

Scheme 1.2

Reversible reactions with proteins, deactivation of enzymes, and finally tumour-inhibiting effects, which are primarily caused by inhibition of protein biosynthesis and of amino acid–RNA-synthetases, are based on these reactions (a) and (b).

For the 2-alkenals the following reactions are known:

(a) Reactions with the sulphydryl group, for which primary addition to the α,β-olefinic group occurs exclusively. Since the velocity constant of the reaction increases by a factor of ten, if the pH value increases by one

Introduction

unit, it can safely be assumed that the reactive group is the sulphide anion, which undergoes reaction with the polarized ethylenic group (scheme 1.3).

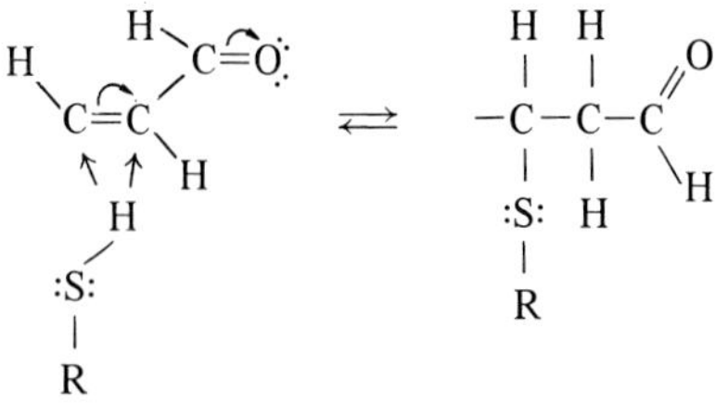

$$\underset{\underset{\underset{R}{|}}{:S:}}{\overset{H}{\underset{|}{C}}}\overset{H}{\underset{\uparrow}{\underset{H}{C}}}\overset{H}{\underset{\nwarrow}{C}}\overset{H}{\underset{}{\overset{}{=}}}\overset{}{\ddot{O}}: \quad \rightleftharpoons \quad -\underset{}{\overset{H}{\underset{|}{C}}}-\underset{\underset{R}{|}}{\overset{H}{\underset{|}{\underset{:S:}{C}}}}-\overset{}{\underset{H}{\overset{O}{\underset{}{C}}}}\overset{\nearrow}{}$$

Scheme 1.3

If the reactant is cysteine, and is present in excess, the initially formed 1:1 adduct immediately reacts with a second molecule of cysteine at the aldehyde group and forms a thiazolidinecarboxylic acid. The reaction occurs so rapidly that it is not possible to trap the 1:1 adduct (scheme 1.4).

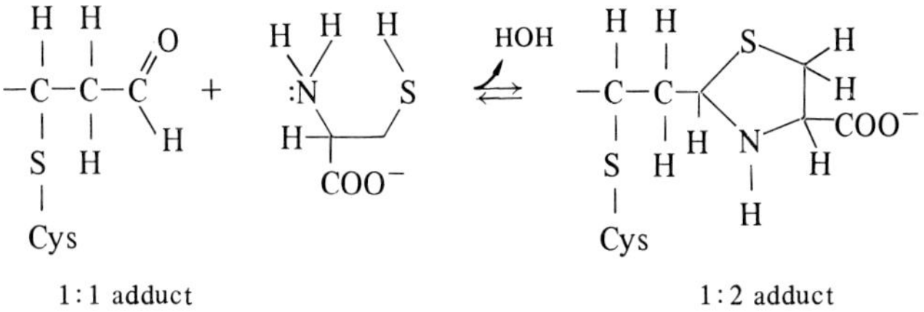

$$\underset{\text{Cys}}{\overset{H}{\underset{|}{C}}-\overset{H}{\underset{|}{C}}-\overset{O}{\underset{H}{C}}\,}_{\underset{|}{S}\;H}^{} \;+\; \underset{}{\overset{H}{\underset{H-}{\overset{}{N}}}\underset{}{\overset{H}{\underset{}{\underset{COO^-}{\overset{}{\diagdown}}}}{S}}} \;\overset{HOH}{\rightleftharpoons}\; \underset{\text{Cys}}{\overset{H}{\underset{|}{C}}-\overset{H}{\underset{|}{C}}}\underset{S\;H}{\overset{}{\underset{}{}}}\;\underset{H}{\overset{S}{\underset{N}{\diagup\diagdown}}}\underset{H}{\overset{H}{\underset{H}{}}}\;\overset{COO^-}{}$$

1:1 adduct   1:2 adduct

Scheme 1.4

(b) Reactions with amino groups, where the formation of Schiff's bases as well as 1,4-addition products is possible. It is generally true, however, that the reaction with the sulphydryl groups takes place much faster, in fact by several orders of magnitude.

It is also certain that the reversible reactions of 2-alkenals with proteins are based on reaction (a) and probably also on reaction (b), and that they may lead to the deactivation of different enzymes—in particular of sulphydryl enzymes—and thus the tumour-inhibiting effects of 2-alkenals can, at least in part, be explained. The reactions referred to are similarly involved in the interpretation of the deactivation of viruses by 2-alkenals, because they explain the finding that 2-alkenals react directly with the protein component of the virus particle, thereby deactivating it.

On the other hand, for the 4-hydroxy-2-alkenals, a subgroup referred to several times above, only the reaction with the sulphydryl group could be established. As with the 2-alkenals it is virtually exclusive addition to the ethylenic group which initially occurs. The saturated aldehyde which is thus formed, with a thioether grouping at C-3, readily undergoes an intramolecular cyclization reaction, involving the hydroxyl group at C-4, to give a hemiacetal.

A special case again is the reaction involving cysteine. With an equimolar mixture of reactants the 1:1 adduct is formed exclusively (scheme 1.5).

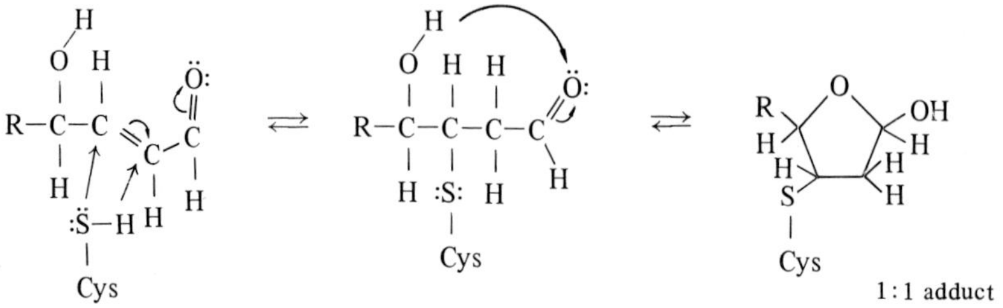

Scheme 1.5

With appropriate excess of cysteine, the initial adduct can combine with another molecule of the former to give a 1:2 adduct, with formation of a thiazolidinecarboxylic acid at the aldehyde function (scheme 1.6).

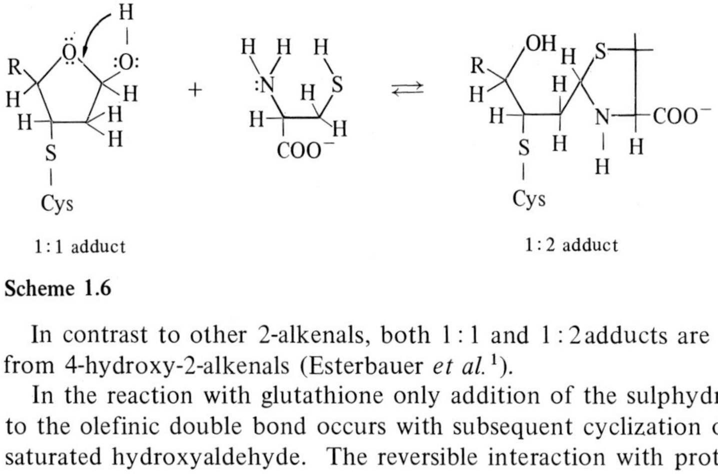

Scheme 1.6

In contrast to other 2-alkenals, both 1:1 and 1:2 adducts are obtainable from 4-hydroxy-2-alkenals (Esterbauer et al.[1]).

In the reaction with glutathione only addition of the sulphydryl group to the olefinic double bond occurs with subsequent cyclization of the saturated hydroxyaldehyde. The reversible interaction with proteins also depends on the addition of sulphydryl groups to the carbon–carbon double bond at C-2 and C-3. This leads particularly to deactivation of sulphydryl-functional enzymes, and with that to the inhibition of nucleic acid and protein biosyntheses, and finally also to an inhibition of the growth of tumours and the multiplication of *Vaccinia* virus. In essential contrast to all other aldehydes no direct reaction with nucleic acids could be established. Hydroxyenals also have a special position in the inhibition of multiplication of the *Vaccinia* virus in that, again in fundamental contrast to α-ketoaldehydes (and possibly also to the 2-alkenals), they do not themselves attack and deactivate the virus particle but solely inhibit sulphydryl enzymes, which are of functional significance for virus multiplication in the infected cell.

An unsaturated group is also in conjugation with an aldehyde group in the α-ketoaldehydes. In contrast to alkenals and hydroxyalkenals, however, the α-keto group here activates the aldehyde group, so that the latter, as with alkanals, becomes the group which reacts primarily. Biological and biochemical inhibitions depend on a reaction of the aldehyde group with the sulphydryl group of cysteine and glutathione. With the former, thiazolidinecarboxylic acids, and with the latter, a hemimercaptal, are formed. Furthermore with the amino groups, especially of arginine and guanine, secondary reactions of the keto group lead to the formation of stable five-membered rings and special effects of α-ketoaldehydes on proteins (enzymes) and nucleic acids depend on this mechanism. Tumour-inhibiting effects are brought about mainly by inhibitions of protein and of DNA biosyntheses. Inhibitory effects on the multiplication of viruses, however, involve direct chemical reaction with the protein, as well as the nucleic acid, components of the virus particle.

1.6 What will the future bring?

Having surveyed in general terms the biological research on aldehydes and its historical development up to the present, let us cast a brief look into the future.

The main problem of general significance perhaps remains the certain and unequivocal experimental proof that certain aldehydes recognized as highly reactive compounds (α-ketoaldehydes, α,β-unsaturated aldehydes) are in fact essential bioregulators of metabolism.

Closely connected with this, there are the following questions: are such aldehydes universal and ubiquitous components of living materials (as some findings appear to indicate)? The chief difficulty confronting this type of investigation is due in particular to the instability of such substances, and to their short lifetime and occurrence in minimal concentrations.

How do such aldehydes arise *in vivo*, and what are the precise mechanisms of their actions?

It is from such investigations that one may expect to derive new weapons for the fight against the growth of malignant tumours and the multiplication of viruses and bacteria.

Problems of more special significance. The special functions of aldehydes in relation to the following phenomena can be said to be particularly interesting:
The transmission of signals in the animal kingdom;
the protective action for plants against bacteria and fungi;
bioluminescence and the visual process.

Finally, we should point out the significance of biological research on aldehydes to problems of environmental pollution, and of food technology [food preservation, the spoilage of fats and oils (rancidity), the spoilage of beer (staleness), and the presence of fragrant components in food and drink].

From investigations of this type one can expect to obtain new insight into the admissibility of using high-energy radiation for food preservation, and into the production and storage of fat-containing foodstuffs and of different beverages.

A review kept along general lines, of course, readily falls into the danger of saying less and less about more and more, until finally, according to the well-known *bon mot*, one says Nothing about Everything. This seems a good time, therefore, to close this chapter and to get on with the more detailed work.

Reference
1 Esterbauer, H., Ertl, A., Scholz, N., 1976, *Tetrahedron*, **32**, 285.

Part 1

Schematic survey of the most important aldehydes involved in intermediary metabolism

Saturated aldehydes

2.1 Structure of saturated aldehydes in water

Saturated aldehydes tend to be hydrated in aqueous solution. The degree of hydration may be determined by Raman spectroscopy[3], or by absorption spectroscopy from the intensity of the carbonyl absorption band in the region of 280 nm[1] or at 5·75 μm [2], and also by chemical methods[4].

Formaldehyde in concentrations up to approximately 1 M is present in aqueous solution mainly or exclusively as the hydrate, methylene glycol [$CH_2(OH)_2$]. With higher aldehydes the degree of hydration rapidly decreases with increasing chain-length and increased extent of branching. α-Ketoaldehydes, such as glyoxal and methylglyoxal, as well as α-hydroxyaldehydes, are always strongly hydrated, whereas the degree of hydration with α,β-unsaturated aldehydes is vanishingly small[4,5]. Some equilibrium constants for the hydration are given in table 2.1. In certain reactions aldehydes react only in the hydrated form, in others only in the free form. Thus Milch[2] has shown that it is the hydrated aldehyde in particular which produces cross-linkages in proteins. In contrast the free aldehyde is the substrate for xanthine oxidase[6] and for aldehyde dehydrogenases from yeast[7] or liver[5].

Table 2.1. Degree of hydration and hydrate dissociation constants of some aldehydes at 25°C at pH 5·0. $K = [RCH(OH)_2]/[RCHO]$. Values partly from Le Henaff[4].

Aldehyde	Hydration (%)	K
formaldehyde	>99·8	>100
acetaldehyde	49·7	0·99
propionaldehyde	47·1	0·89
butyraldehyde	39·5	0·65
isobutyraldehyde	38·2	0·61
trimethylacetaldehyde	19·0	0·23
acrolein	<2·0	<0·02
crotonaldehyde	<2·0	<0·02

2.2 Reactions of alkanals with amino acids

The most reactive aldehyde of the alkanal series is formaldehyde. A very large number of investigations have been made into the reactions of this aldehyde with amino acids, peptides, and proteins. French and Edsall[8] have written a review article on this topic. Many of the reactions described require extreme pH values, a raised temperature, high concentrations of reactants, special solvents, etc. It is therefore questionable whether many of these reactions can occur under physiological conditions.

In what follows a few of the most important reactions of aldehydes are briefly discussed, and may possibly contribute to the explanation of some of their biological effects.

The uncharged amino group of monofunctional amino acids (amino acid anion) combines rapidly and reversibly with one or two molecules of formaldehyde (probably in the form of the hydrate), N-hydroxymethyl-amino acids and NN-bis(hydroxymethyl)amino acids[9,10] (cf scheme 2.1) being formed.

$$\begin{array}{cccc} \text{RCHCOO}^- & \text{RCHOO}^- & \text{RCHCOO}^- & \text{RCHCOO}^- \\ | & \rightleftharpoons \; | & \rightleftharpoons \; | & \rightleftharpoons \; | \\ ^+\text{NH}_3 & \text{NH}_2 & \text{NHCH}_2\text{OH} & \text{N(CH}_2\text{OH)}_2 \end{array}$$

Scheme 2.1

The basicity of the amino group of the hydroxymethylamino acids is appreciably lower than that of the unsubstituted amino acids. In the presence of 9% formaldehyde the pK_2 value of glycine is shifted from 9·60 to 5·92[11]. Formaldehyde added to a neutral solution of an amino acid strongly decreases the basicity of the amino group, but does not alter the amount of alkali needed. This principle is utilized in the well known formol titration procedure of Sörensen[12,13].

The association constants for hydroxymethylamino acids (AF^-) $L_1 = [AF^-]/[A^-][F]$ (where A^- is the amino acid anion and F is formaldehyde), lie in the range twenty to one hundred. The stability of the hydroxymethyl compounds is thus not specially great, and only with high concentrations (10^{-1} M - 10 M) of formaldehyde does the equilibrium lie appreciably over to the side of the hydroxymethyl compounds. On dilution of the solution immediate dissociation takes place. Since the uncharged amino group is the reactive form it reacts the more readily with formaldehyde the lower its basicity.

The higher alkanals (acetaldehyde, etc.) react with the amino groups of amino acids and primary amines to give carbinolamines and Schiff's bases respectively in readily reversible reactions:

$$\text{RCHO} + \text{H}_2\text{NR}' \rightleftharpoons \text{RCH(OH)NHR}' \rightleftharpoons \text{RCH} = \text{NR}' + \text{H}_2\text{O}$$

In aqueous solution and at neutral pH a very large excess of free aldehyde is required to shift the equilibrium in favour of the Schiff's base[14-17].

If in addition to the amino group other functional groups are also present in the amino acid or peptide, the number of possible reactions increases. As a rule stable heterocyclic products result. Once again the reactions of formaldehyde have been well studied, and these are summarized in scheme 2.2.

Formaldehyde also reacts very rapidly with tetrahydrofolic acid in aqueous solution to give N^5, N^{10}-methylenetetrahydrofolic acid, which may then act as the substrate for enzyme-catalysed hydroxymethylations and methylations[18].

With two molecules of adrenaline or noradrenaline formaldehyde gives a reversible complex. This reaction is regarded as the model for the reversible interaction of catecholamines and their receptors (α-adreno-receptors). It is assumed that the receptor involves an aldehyde group in the active centre[19].

The only reaction shown in scheme 2.2 which readily takes place with longer chain aldehydes is the reaction with cysteine to give thiazolidine-carboxylic acids[20, 21]. The formation of these acids in tissues and tissue extracts in the presence of exogenous aldehydes has recently been investigated by Loreti et al.[22]. These authors believe that the inhibitory effect of some aliphatic aldehydes on protein biosynthesis is based on the lowering of the cysteine concentration through the formation of thiazolidinecarboxylic acid[23].

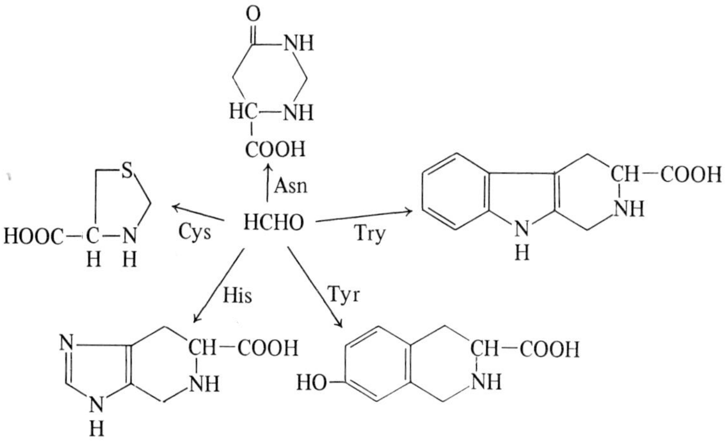

Scheme 2.2

2.3 Effects of alkanals on proteins

The reaction of formaldehyde with proteins has been investigated by many research workers[8, 15, 16, 24-32]. Used in high concentrations with proteins, formaldehyde produces inter- and intra-molecular cross-linking. This process is of technical importance in tanning procedures. It is also of biochemical interest since formaldehyde, as a consequence of the above properties, is very commonly used as a fixative in cytochemical and electron-microscopic examinations of biological structures (reviews by Sabatini et al.[33], Reale and Luciano[34]). Other saturated monoaldehydes which exhibit this protein cross-linking, either not at all or only to a small extent, have hardly been investigated so far. Milch, however, in a series of investigations, has examined other aldehydes, in particular those possibly involved as intermediates in metabolic processes[15]. In the opinion of Milch, collagen-reactive aldehydes play an important role in the biological ageing process of connective tissues. Despite many investigations the chemical composition of the products of reaction between aldehydes and proteins is not yet precisely established. In most cases two types of reaction product could be distinguished, the first being unstable, that is, more or less easily dissociable, and the second stable and not dissociable. Several aldehydes, especially formaldehyde and bifunctional aldehydes such as glyoxal, glutaraldehyde, etc., give both types of reaction product. Most other aldehydes, however, such as acetaldehyde, propionaldehyde, etc., may only be bound reversibly to proteins.

The amino groups of the side chains of lysine and arginine are primarily responsible for the reversible binding of aldehydes to proteins. Aldehydes initially give the very unstable carbinolamine (see section 1.2), and these can then give the somewhat more stable Schiff's base through subsequent dehydration (cf scheme 2.3).

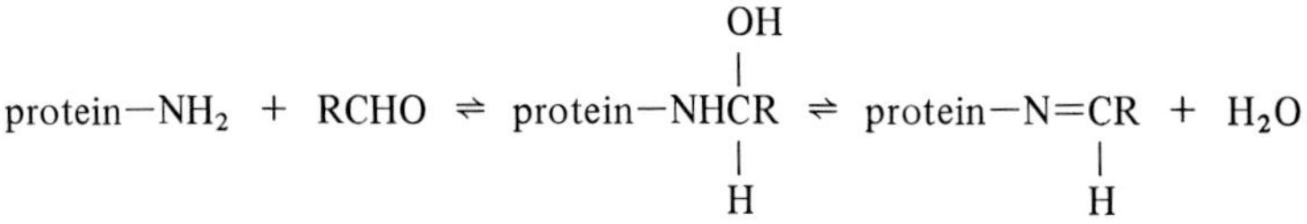

$$\text{protein}-NH_2 + RCHO \rightleftharpoons \text{protein}-\underset{H}{\underset{|}{N}}H\underset{}{\overset{OH}{\underset{|}{C}}}R \rightleftharpoons \text{protein}-N=\underset{H}{\underset{|}{C}}R + H_2O$$

Scheme 2.3

It is probable that aldehydes may be reversibly bound not only at the amino group but at other functional groups of serine, cysteine, histidine, tyrosine, and tryptophan and to some extent also at the amide nitrogen group (NH) of the peptide linkage. The free amino groups are again chiefly responsible for the irreversible binding of aldehydes to proteins. Aminomethylols are mostly formed in an equilibrium reaction, and the very reactive methylol group then undergoes further reaction with a second reactive grouping of another protein molecule. This involves dehydration, and a methylene bridge is introduced as an intermolecular cross-linkage. According to Milch, only those aldehydes which are present predominantly as hydrates have cross-linking properties[2] (cf scheme 2.4).

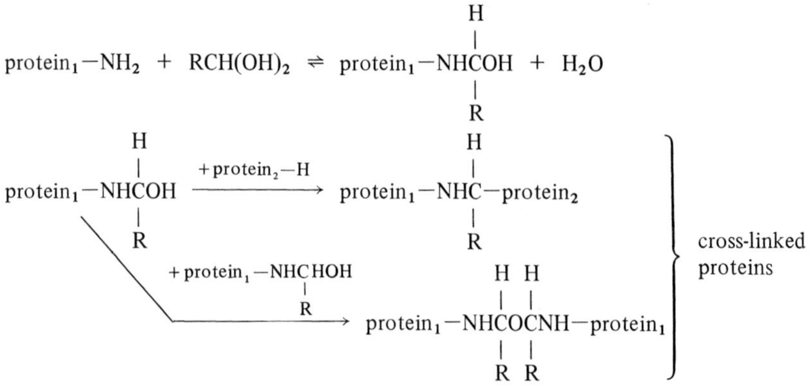

$$\text{protein}_1-NH_2 + RCH(OH)_2 \rightleftharpoons \text{protein}_1-\underset{R}{\underset{|}{N}}H\underset{}{\overset{H}{\underset{|}{C}}}OH + H_2O$$

$$\text{protein}_1-\underset{R}{\underset{|}{N}}H\underset{}{\overset{H}{\underset{|}{C}}}OH \xrightarrow{+\text{protein}_2,-H} \text{protein}_1-\underset{R}{\underset{|}{N}}H\underset{}{\overset{H}{\underset{|}{C}}}-\text{protein}_2$$

$$\xrightarrow{+\text{protein}_1-NHCHOH \atop R} \text{protein}_1-NHC\underset{R\ R}{\overset{H\ H}{\underset{|\ |}{O}}}CNH-\text{protein}_1$$

} cross-linked proteins

Scheme 2.4

The reactive group for the second binding site may be an amide, guanidyl, indolyl, hydroxyphenyl, imidazolyl, hydroxyl, or sulphydryl group. Free amino groups do not appear to be significant here. There is also evidence that ether bridges (>CH—O—CH<) are in part present, being formed by the condensation of two aminomethylol groupings.

Indirect as well as direct methods have been employed for the determination of the extent of aldehyde binding of proteins. Indirect methods comprise: recording of titration curves, determination of the amino groups converted, measurement of UV spectra, viscosity, swelling, temperature of shrinkage, denaturation, and estimation of molecular weight, osmotic pressure, and dielectric properties. For the direct determination of formaldehyde, hydrolysis of the protein with hot acid is recommended; the formaldehyde formed is carefully distilled off and determined in the distillate either with dimedone or titrimetrically (cf references 8 and 16).

One problem in the determination of protein-bound aldehydes is that a different aldehyde content may be obtained according to the method of preparation, owing to variations in the stability of the binding. Nitschmann and Hadorn[25] showed, for example, with formaldehyde-treated casein, that a part of the formaldehyde is only very weakly bound, and may be removed by a short washing procedure. Further washing over several days, or alternatively lengthy dialysis, removes a further part of the aldehyde from the protein; a remaining part, however, is so firmly bound that dialysis carried out over several weeks produces no appreciable dissociation. Even with the splitting off of bound aldehyde through acid hydrolysis the reaction conditions will determine which kind of linkage is split. Thus, aldehyde bound in the form of a Schiff's base may be split off from protein by means of a weak acid or even with hot water. The expression 'protein-bound aldehyde' should therefore be defined very carefully and with reference to the conditions of preparation.

2.3.1 *Special proteins: collagen*
The use of formaldehyde in the manufacture of leather has provided stimulus for many theoretical studies[16, 24, 30, 31, 35]. Collagen treated with formaldehyde is inert towards proteolytic enzymes. Its swelling in water, acid, or alkali decreases. The temperature of thermal shrinkage is raised from 75°C to 90°C, and the shrinkage is reversible, in distinction to that found for natural collagen. The quantity of bound formaldehyde rises steeply with increasing pH; it is supposed that up to pH 7–8 it is the amino groups of lysine which mainly react, and that above pH 8 the guanidine groups of arginine will also react.

The alteration of collagen structure is one of the most characteristic changes that may be shown to occur in pathologically or naturally aged connective tissue. There exist many analogies between collagen 'aged' *in vivo*, and collagen treated *in vitro* with cross-linking aldehydes (formaldehyde, dialdehydes, α,β-unsaturated aldehydes). Milch[15] assumes that agents of metabolic origin, which are capable of cross-linking collagen, accumulate in connective tissue with increasing age, and finally alter the structure of the collagen by developing such cross-linkages. Many investigations[2,15,36-40] have been reported on the influence of normally present products of intermediary metabolism on various collagen and gelatin preparations. The results indicate that water-soluble aldehydes of lower molecular weight (formaldehyde, glyoxal, glycolaldehyde, glyceraldehyde, acrolein, crotonaldehyde) combine with the protein chains of collagen by cross-linking them. The proteins thus modified by aldehydes show many of the properties of collagen in aged connective tissue.

It follows from the researches of Cater[16,41] that different aldehydes produce very different types of cross-linkages in collagen. Thus formaldehyde, acrolein, succinaldehyde, glutaraldehyde, and glyoxal give relatively stable cross-linkages, whereas propionaldehyde, methylglyoxal and adipaldehyde do not introduce any cross-linkages into the collagen molecule.

2.3.2 *Special proteins: serum albumin*
If a 10% solution of serum albumin is treated at pH 3·5-7·5 with formaldehyde (0·4-10%) intermolecular cross-linkages are introduced. Through this attachment of several protein molecules protein polymers are obtained with two to five times the molecular weight. The existence of polymers was proved by the lowering of osmotic pressure[28] and by a decrease in the elution volume on separations with Sephadex G200[42]. Serum albumin treated with formaldehyde shows a decrease, but no qualitative alteration, of its antigenic properties. This observation may be due to part of the antigen-determining positions within the protein molecule being displaced through inter- and intramolecular cross-linking[42,43]. Treatment of serum albumin with glutaraldehyde, however, leads to the creation of an additional antigenic region[42] (cf section 5.3).

Since all aldehydes attack mainly the basic amino group in proteins, it is clear that treatment with aldehydes (formaldehyde, glutaraldehyde, α-hydroxyadipaldehyde) will lower the isoelectric point and increase electrophoretic mobility in the alkaline region[42]. Okulov[44] has studied the influence of a series of aldehydes (formaldehyde, acetaldehyde, propionaldehyde, butyraldehyde, isovaleraldehyde, citral, benzaldehyde, furfural) on the electrophoretic mobility of serum, serum albumin, and γ-globulin. According to his findings albumin is less susceptible than γ-globulin, and short chain aldehydes alter the electrophoretic properties more than do those of longer chain-length.

2.3.3 *Special proteins: lipase*
As early as 1935 Weinstein and Wynne[45] had reported that formaldehyde, acetaldehyde, butyraldehyde, aldol, and benzaldehyde inhibited pancreatic lipase. Monty[46] suggests that the inhibition of lipase caused by decanal is

not based on a reaction of the aldehyde with reactive groups of the enzyme, but on an alteration of the emulsoid state of the fat-droplets which serve as substrate. Landsberg and Sinnhuber[47] have examined the inhibition of lipase *in vitro* by malonaldehyde, formaldehyde, and propionaldehyde, malonaldehyde giving the greatest inhibition. Formaldehyde is appreciably less effective, concentrations of 10^{-2} M having practically no influence on the activity. Traces of methanol, however, strongly enhance the efficiency of the formaldehyde. Propionaldehyde initially causes a slight inhibition at 10^{-1} M, but after further incubation this completely disappears. Since aldehydes are generally protein-reactive compounds that are often present in foodstuffs, the influence of aldehydes on the enzymes of digestion is of physiological importance.

2.4 Some physiological effects of alkanals

Serotonin metabolism is influenced in a remarkable way by saturated aldehydes (cf section 8.2). Serotonin has a very rapid metabolic turnover. Either it is oxidized by a monoamine oxidase to 5-hydroxyindol-3-ylacetaldehyde, which in its turn is oxidized to 5-hydroxyindol-3-ylacetic acid, or it is converted into the 5-hydroxyglucuronide or the 5-hydroxy sulphate. It was subsequently shown that formaldehyde, acetaldehyde, and propionaldehyde inhibit not only oxidative deamination of serotonin[48,49], but also the enzymatic oxidation of hydroxyindol-3-ylacetaldehyde to the acid[50].

These inhibitory effects of aldehydes produce accumulation of serotonin in perfused frog liver, and a displacement of the main metabolic path (oxidative deamination) in the direction of conjugation (formation of sulphate and glucuronide)[51]. James and Bear[52] have reported that a series of aldehydes (acetaldehyde, propionaldehyde, butyraldehyde) have a sympathomimetic action on the heart and the heart muscle. The probable explanation is a raised serotonin concentration caused by these aldehydes, which induces a sympathetic stimulus similar to that caused by adrenaline.

There is the interesting observation[53] that short-chain aldehydes, like formaldehyde, acetaldehyde, propionaldehyde, and acrolein, induce a release of histamine in lung slices (guinea pig). Acetaldehyde is the most active and propionaldehyde the least active. It was supposed that acetaldehyde, which occurs in cigarette smoke, is responsible for the release of histamine caused by smoking. Several toxicological studies[54-56] are concerned with the pathological and physiological characterization of the different effects shown by short-chain aldehydes that are present in cigarette smoke, in exhaust gases, and in the atmosphere. Acute toxicity decreases in the sequence acrolein, crotonaldehyde, acetaldehyde, butyraldehyde, isobutyraldehyde, valeraldehyde. All aldehydes are highly irritant substances and cause pathological changes, mainly in the respiratory tract, in particular in the mucosa.

2.5 Effect of alkanals on cell respiration
2.5.1 *Formaldehyde*
Formaldehyde inhibits the oxidation of succinate, glutamate, α-oxoglutarate, and pyruvate, which are coupled with phosphorylation. Uncouplers may restore the respiration[88]. NADH-oxidation of submitochondrial particles is not affected. Formaldehyde similarly inhibits the dinitrophenol-stimulated adenosine triphosphatase (ATPase) and the ATP-dependent reversal of electron transport in submitochondrial particles. The inhibition of the reversal of the electron transport may be partially reversed by increasing the ATP concentration. Tyler[89] assumes that the effects of formaldehyde are based on inhibition of transport of inorganic phosphate and on attack on the NAD-flavine region of the respiration chain. On the other hand, Van Buskirk and Frisell[88] believe that formaldehyde interferes with the energy transfer somewhere between the site of coupling with the electron transport and the site of ATP synthesis.

Since formaldehyde would be able to develop its effect in a reaction with sulphydryl groups, the interpretation by Tyler appears to be the more likely one to the authors, since electron transport in the NAD-flavine region[104] as well as the translocation of inorganic phosphate across the inner membrane of mitochondria[105-107,89] involves sulphydryl groups.

2.5.2 *Other aldehydes*
Acetaldehyde, propionaldehyde, isobutyraldehyde, glyoxalate, isovaleraldehyde, methylglyoxal, and succinaldehydic acid exhibit effects on the oxidation processes in the mitochondrion[90-93]. Of all these aldehydes methylglyoxal and isovaleraldehyde are very active inhibitors of pyruvate metabolism. Succinaldehydic acid and glyoxalate affect pyruvate oxidation only at high concentrations. Of the monofunctional saturated aldehydes isovaleraldehyde is the most effective. Acetaldehyde, propionaldehyde, isobutyraldehyde, and isovaleraldehyde are oxidatively metabolized by the mitochondria. The appropriate dehydrogenase shows the lowest activity with isovaleraldehyde and the highest with acetaldehyde. Propionaldehyde and isovaleraldehyde compete with acetaldehyde for the enzyme; isovaleraldehyde is the more effective inhibitor of acetaldehyde oxidation.

Acetaldehyde inhibits the respiration of pyruvate in mitochondria from liver, kidney, cerebrum, muscle, and cerebellum, in that order of increasing effectiveness. The low sensitivity of liver and kidney mitochondria was thought to be due to their capacity to oxidize acetaldehyde[92,93]. After contact with acetaldehyde only mitochondria from liver and kidney recover quickly.

NAD can in part reverse the inhibition of intact mitochondria. A complete restitution of respiration is achieved only with inhibition of glutamate metabolism, but only if the mitochondria have become permeable to NAD through ageing. The NADH oxidation in mitochondria, which have been made permeable for this coenzyme, is not impaired by the aldehydes examined. From these findings it was concluded that an attack occurs between the entry of the substrate and the NADH dehydrogenase[95]. Little or no inhibition of oxidation was found for α-glycerophosphate, succinate, and β-hydroxybutyrate[94].

Higher alcohols, the so-called fusel oils, are present in nearly all alcoholic drinks. Oxidation of these alcohols leads to the corresponding aldehydes, which then, like the aldehydes mentioned above, inhibit oxidation of pyruvate and acetaldehyde, and thereby inhibit the metabolic pathways of ethanol and its metabolites. One of the effects of these aldehydes would be the appearance of a 'hangover' after excessive consumption of alcohol[94].

2.6 Antitumour effects of alkanals

Between 1932 and 1939 Strong[58-62] reported a series of investigations on the influence of essential oils on tumour growth in mice. When oil of wintergreen or oil of thyme was added to the feed, tumours developed less frequently and more slowly in a breed of mice that were susceptible to breast tumours. Further experiments showed that only the lower boiling point fractions of these essential oils had antitumour activity, and the main portion of the higher boiling point substances had no effect. From the lower boiling point fractions the active substance heptanal (oenanthal) was isolated. Strong was also able to show that heptanal is not only prophylactically active against breast tumours, but also inhibits the further growth of spontaneous tumours already present. In some instances the tumour underwent a softening and a liquefaction, leaving behind a haemorrhagic zone which was filled with a bloody fluid. If heptanal was injected subcutaneously in the vicinity of the tumour, liquefaction again took place.

The experiments of Strong were repeated and extended by Boyland[63] and Dittmar[64]. Dittmar inoculated mice with Ehrlich carcinoma. Three days later he repeatedly administered heptanal, diluted with olive oil (0.1 ml of a 5% solution), subcutaneously on the side opposite the tumour, and likewise observed a regression of the tumour. Citronellal and citral acted similarly to heptanal. Dittmar believes that tumour inhibition depends on capillary damage caused by the aldehydes.

Boyland[63] and coworkers examined the effect of a series of aldehydes and ketones on the growth of spontaneous breast tumours and inoculated tumours (sarcoma-180). They found that heptanal and citral effectively inhibited in particular the growth of spontaneous tumours in mice. It was concluded that heptanal could possibly be converted by oxidative degradation into pimelic acid, glutaric acid, and malonic acid, and that these acids were the actual cause of the tumour inhibition. These dicarboxylic acids, especially malonic acid, do in fact exert a pronounced inhibitory effect on the growth of tumours.

A number of more recent investigations have been concerned with the mechanism of tumour inhibition by aliphatic aldehydes, and the relationship between chemical constitution and antitumour activity[22,23,65-69].

Some of the results of these investigations are summarized in table 2.2. It may be seen that the aldehydes investigated do not measurably, or at most only slightly, influence the oxygen consumption of the liver and the hepatoma. In contrast all the aldehydes studied appreciably inhibit protein biosynthesis, as shown by the lowered incorporation of leucine into liver and tumour protein. This inhibition is related to the presence of a reactive aldehyde group, since the corresponding acids (for example, isobutyric acid, α-hydroxybutyric acid, or β-hydroxybutyric acid) in equal concentration do not affect protein synthesis. In addition, the molecular structure of the aldehyde is significant for the extent of inhibition in normal tissue and in tumour tissue. As table 2.2 indicates, some aldehydes (propionaldehyde, butyraldehyde, isobutyraldehyde, crotonaldehyde, lactaldehyde) act preferentially on liver, whereas others (α-hydroxybutyraldehyde, β-hydroxybutyraldehyde, and all four stereoisomeric α,β-dihydroxybutyraldehydes) preferentially inhibit protein synthesis in the tumour. It was concluded that all the aldehydes examined intercept intracellular cysteine through formation of the corresponding thiazolidinecarboxylic acids (scheme 2.2), and thereby disturb the balanced amino acid pool necessary for protein synthesis.

The primary cause of the inhibition of protein synthesis would then be the reduced availability of cysteine. The different susceptibilities of healthy tissue and of tumour tissue may be explained by the differences in the metabolism of the thiazolidinecarboxylic acids. It is known, for

Table 2.2. Effects of aliphatic aldehydes on oxygen consumption and on leucine incorporation into protein with liver slices and Yoshida AH 130 ascites hepatoma cells *in vitro*. The aldehyde concentration was 5 mM. Values were taken from Perin *et al.*[69].

Aldehyde	Inhibition (%)			
	liver slice		hepatoma cells	
	oxygen consumption	leucine incorporation	oxygen consumption	leucine incorporation
acetaldehyde	—	—	0·2	51
propionaldehyde	0	78	—	37
butyraldehyde	0	87	0	38
isobutyraldehyde	4	98	0	48
crotonaldehyde	5	58	0	31
lactaldehyde	14	88	7	66
glyceraldehyde	11	77	15	57
α-hydroxybutyraldehyde	0	36	12	75
β-hydroxybutyraldehyde	25	78	36	99
α,β-dihydroxybutyraldehyde	10	55	5	91

example, that in isolated mitochondria as well as in the living animal, thiazolidinecarboxylic acid is metabolized by specific dehydrogenases[70]. It is possible that neoplastic tissue may be different from normal tissue in this respect[23].

Another possible explanation for the inhibitory action of aldehydes on protein biosynthesis could be found in the fact that aldehydes inhibit the leucyl-RNA synthetase. The inhibitory effect decreases in the sequence formaldehyde, glyceraldehyde, glyoxal, glyoxalate, glycolaldehyde, and acetaldehyde[71].

In conclusion the researches of Weitzel[72, 96-103] and coworkers, who have prepared numerous compounds which inhibit the growth of the Ehrlich ascites tumour in the mouse, and also in part that of the Walker carcinosarcoma in the rat, should be mentioned. The cytostatic agent Natulan [4-(2-methylhydrazino)methyl-N-isopropylbenzamide], introduced into therapy in 1963, is also effective in the treatment of human tumours. Scheme 2.5 summarizes the classes of compounds examined by these authors. Common to all five classes is the ready formation of formaldehyde *in vitro*. It was thus obvious to assume that the tumour-inhibiting action of these compounds was at least in part due to the cytotoxically active formaldehyde being set free intracellularly.

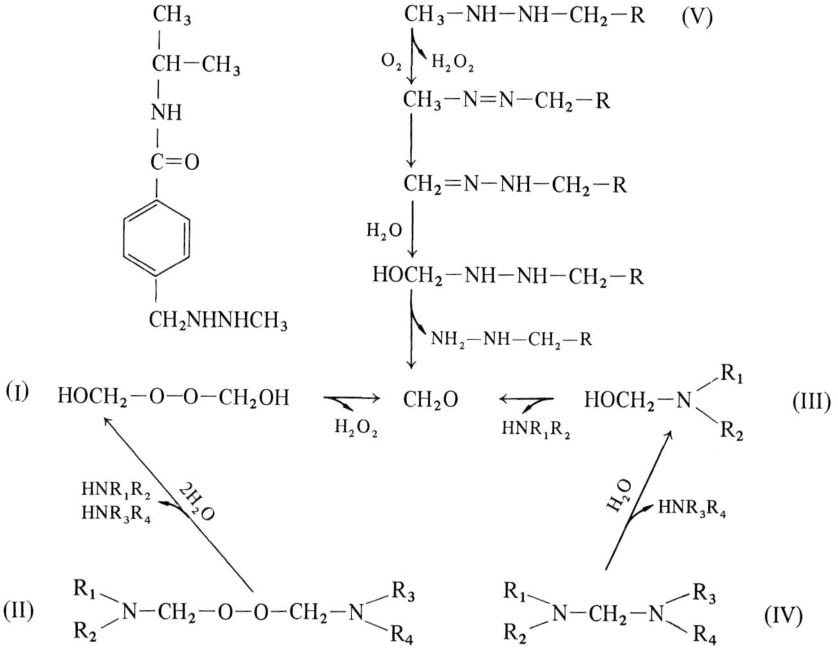

Scheme 2.5

This assumption is supported by the finding that formaldehyde itself has tumour-inhibiting activity. Thus, by means of doses of 1 mg per day per mouse, a total inhibition of the growth of Ehrlich ascites tumour can be achieved. From the fact that some compounds of types (I)–(V) (scheme 2.5) remain active as tumour inhibitors in concentrations at which formaldehyde itself exhibits no more activity, the authors conclude that several factors are involved in the mechanism of action, namely:
(a) hydrogen peroxide is set free intracellularly (with types I, II, V);
(b) formaldehyde is set free intracellularly (with types I–V);
(c) the alkylating action of the $-CH_2OH$ group (with types I–V);
(d) the alkylating action of the $>N-CH_2-N<$ group (with type IV);
(e) from Natulan, through degradation of the side-chain group R, certain active intermediate products may be formed *in vivo*, which possess an aldehyde group, for example, 4-formyl-*N*-isopropylbenzamide.

2.7 Effects of alkanals on bacteria and fungi

Formaldehyde has often been used for the deactivation of bacterial toxins, as required for the preparation of vaccines. In general a formaldehyde concentration of 0·12–0·16% and an incubation period of several weeks at 37°–39°C is recommended. The reaction therefore is very slow and practically irreversible. In these circumstances the reactions which take place have hardly been investigated at all; certainly, however, one is not dealing simply with reactions of the formaldehyde with amino groups as in the formol titration method. Toxins treated with formaldehyde retain their immunological specificity as antigens to a considerable extent.

Formaldehyde has been used for the production of diphtheria and tetanus toxoids[73-75], snake venom toxoids[76,77], and plant toxoids[78].

Rosenkranz[108] reported that formaldehyde preferentially inhibits the growth of a strain of *Escherichia coli* which lacks the enzymes involved in DNA repair (pol A$^-$ strain). It was assumed that the DNA of the cells was altered by the reaction of formaldehyde with the amino groups of the DNA bases[109,110]. As preferential inhibition of the pol A$^-$ strain is also exhibited by the well-known carcinogens, methyl methanesulphonate and N-hydroxylaminofluorene, it seems quite reasonable to consider formaldehyde as a possible carcinogen. It can be expected that exposure to formaldehyde is very common, as it is a prominent constituent of polluted water and air[111]. Large quantities are produced from the combustion of gasoline and coal, and formaldehyde comprises approximately 40% to 50% of the total aldehydes present in automobile exhaust gases and in the urban atmosphere. Normally 0·02 to 0·1 ppm of formaldehyde is found in the atmosphere, but values as high as 0·2 ppm have been reported[111]. Formaldehyde is also used widely as a preservative (for example, in dissecting rooms), a germicide and an antimicrobial agent[112,113]. In addition, formaldehyde and formaldehyde releasing compounds (paraformaldehyde, hexamethylene tetramine) are produced in large quantities (in

1963 the USA alone produced 3·5 billion pounds of the 37% aqueous solution of formaldehyde[111]) and used as intermediates in many industrial processes. In view of the findings of Rosenkranz[108] and the fact that formaldehyde is already found throughout the human environment a reevaluation of the continued use of this chemical and its uncontrolled emission to the atmosphere seems to be necessary.

Együd[79] reported that formaldehyde and acetaldehyde at concentrations of 1 mM completely and irreversibly inhibit cell division of *Escherichia coli*. Longer chain aldehydes, however, cause only moderate inhibition, from which the cells quickly recover. Other authors have examined the inhibitory action of formaldehyde on the growth of milk streptococci[80] and of *Staphylococcus aureus*[81]. In the past, concentrated formaldehyde solutions were often used for disinfection purposes; since other chemical disinfectants, in particular glutaraldehyde, act more rapidly and have a wide spectrum of activity, formalin is now employed only for special purposes[82]. A few reports also exist on the effect of long chain aldehydes on fungi and bacteria. Such aldehydes, in particular dodecanal and tridecanal, strongly stimulate light emission *in vitro* by bacterial luciferases[83, 84].

Long chain aldehydes (especially nonanal) stimulate the germination of dormant wheat rust uredospores. It has been assumed that nonanal is the natural endogenous stimulant for the germination, especially since this aldehyde could be isolated from the spores[85].

The growth of fungi (*Hymenomycetes*) is also stimulated by aliphatic aldehydes[85]. Nonanal is the most effective of the alkanals studied, containing five to eleven carbon atoms, and these alkanals act both in the gaseous state and in solution. Investigations by Nyman[86] have shown that the growth stimulating effect can be traced back to a shortening of the resting period (lag phase). The greatest growth stimulating effect is observed when nonanal is added to the nutrient medium at concentrations of 80–160 µM, before the inoculation.

According to Norrman[87] nonanal, nonenal, and acetaldehyde are natural fungal metabolites and generally have a growth promoting influence. Nonenal stimulates general growth in *Pestalotia rhododendri*, and acetaldehyde and nonanal stimulate the formation of mycelia.

References

1. Bloch, C., Rumpf, R., 1953, *C. R. Acad. Sci.*, **237**, 619.
2. Milch, R. A., 1964, *Biochim. Biophys. Acta*, **93**, 45.
3. Kohlrausch, K. W., Köppl, F., 1934, *Z. Phys. Chem. Abt. B*, **24**, 370.
4. Le Henaff, P., 1967, *C. R. Acad. Sci.*, **265**, 175.
5. Bodley, F. H., Blair, A., 1971, *Can. J. Biochem.*, **49**, 1.
6. Fridovich, I., 1966, *J. Biol. Chem.*, **241**, 3126.
7. Naylor, J. F., Fridovich, I., 1968, *J. Biol. Chem.*, **243**, 341.
8. French, D., Edsall, J. T., 1945, *Adv. Protein Chem.*, **2**, 277.
9. Levy, M., 1933, *J. Biol. Chem.*, **99**, 767.
10. Levy, M., Silberman, D. E., 1937, *J. Biol. Chem.*, **118**, 723.

11 Cohn, E. J., Edsall, J. T., 1943, *Proteins, Aminoacids and Peptides* (Reinhold, New York).
12 Sörensen, S. P. L., 1908, *Biochem. Z.*, **7**, 45.
13 Birch, T. W., Harris, L. J., 1930, *Biochem. J.*, **24**, 1080.
14 Layer, R. W., 1962, *Chem. Rev.*, **63**, 489.
15 Milch, R. A., 1963, *Gerontologia*, **7**, 129.
16 Bowes, J. H., Cater, C. W., 1968, *Biochim. Biophys. Acta*, **168**, 341.
17 Koenigstein, J., Fedoronko, M., 1970, *Proc. Anal. Chem. Conf. (Budapest), 3rd*, **2**, 113.
18 Kallen, R. G., 1971, *Methods Enzymol.*, **18**, 705.
19 Manukhin, B. N., Vyazmina, N. M., 1971, *Fiziol. Zh. SSSR im. I. M. Sechenova*, **57**, 372 [*Chem. Abstr.*, **75**, 15726 (1971)].
20 Schuberth, H. P., 1936, *J. Biol. Chem.*, **114**, 341.
21 Ratner, S., Clarke, H. T., 1937, *J. Am. Chem. Soc.*, **59**, 200.
22 Loreti, L., Ferioli, M. E., Gazzola, G. C., Guidotti, G. G., 1971, *Eur. J. Cancer*, **7**, 281.
23 Guidotti, G. G., Loreti, L., Ciaranfi, E., 1965, *Eur. J. Cancer*, **1**, 23.
24 Künzel, A., 1937, *Angew. Chem.*, **50**, 308.
25 Nitschmann, H., Hadorn, H., 1944, *Helv. Chim. Acta*, **27**, 299.
26 Fraenkel-Conrath, H., Olcott, H. S., 1948, *J. Am. Chem. Soc.*, **70**, 2673.
27 Fraenkel-Conrath, H., Olcott, H. S., 1948, *J. Biol. Chem.*, **174**, 827.
28 Fraenkel-Conrath, H., Mecham, D. K., 1949, *J. Biol. Chem.*, **177**, 477.
29 Gustavson, K. H., 1949, *Adv. Protein Chem.*, **5**, 353.
30 Gustavson, K. H., 1956, *The Chemistry of Tanning Processes* (Academic Press, New York).
31 Davis, P., Tabor, B. E., 1963, *J. Polym. Sci. Part A-1*, 799.
32 Lipparini, L., 1967, *Atti Congr. Anal., 6th*, 211 [*Chem. Abstr.*, **74**, 19434 (1970)].
33 Sabatini, D. S., Bensch, K., Barrnett, R. J., 1963, *J. Cell Biol.*, **17**, 19.
34 Reale, E., Luciano, L., 1970, *Histochemie*, **23**, 144.
35 Bose, S. M., Thomas, K., 1957, *J. Am. Leather Chem. Assoc.*, **52**, 200.
36 Milch, R. A., Frisco, L. J., Szymkoviak, E. A., 1965, *Biorheology*, **3**, 9.
37 Milch, R. A., 1965, *South. Med. J.*, **58**, 153.
38 Milch, R. A., 1965, *J. Atheroscler. Res.*, **5**, 215.
39 Milch, R. A., Clifford, R. E., Murray, R. A., 1966, *Nature (London)*, **210**, 1042.
40 Tranavska, Z., Sitaj, S., Gremela, M., Malinsky, J., 1966, *Biochim. Biophys. Acta*, **126**, 373.
41 Cater, C. W., 1965, *J. Soc. Leather Trades' Chem.*, **49**, 455.
42 Hopwood, D., 1969, *Histochemie*, **17**, 151.
43 Habeeb, A. F. S. A., 1959, *J. Immunol.*, **102**, 457.
44 Okulov, V. I., 1963, *Ukr. Biokhem. Zh.*, **35**, 327 [*Chem. Abstr.*, **56**, 11981h; *Chem. Abstr.*, **59**, 10372e].
45 Weinstein, S. S., Wynne, A. M., 1935, *J. Biol. Chem.*, **112**, 641.
46 Monty, K. J., 1960, *Proc. Fed. Am. Soc. Exp. Biol.*, **19**, 1034.
47 Landsberg, J. D., Sinnhuber, R. O., 1965, *J. Am. Oil Chem. Soc.*, **42**, 821.
48 de Breyer, I. J. J., Soehring, K., 1967, *Med. Pharmacol. Exp.*, **17**, 351.
49 de Breyer, I. J. J., Soehring, K., 1968, *Arch. Exp. Pathol. Pharmakol.*, **260**, 148.
50 Lahti, R. A., Majchrowicz, E., 1969, *Biochem. Pharmacol*, **18**, 535.
51 Ortiz, A., Soehring, K., Terke, E., 1971, *Arzneim. Forsch.*, **21**, 116.
52 James, T. N., Bear, E. S., 1968, *J. Pharmacol. Exp. Ther.*, **163**, 300.
53 Saindelle, A., Ruff, F., Flavian, N., Parrot, J. L., 1968, *C. R. Acad. Sci. Ser. D*, **266**, 139.
54 Salem, H., Cullumbine, H., 1960, *Toxicol. Appl. Pharmacol.*, **2**, 183.

55 Murphy, S. D., Davis, H. V., Zaratzian, V. L., 1964, *Toxicol. Appl. Pharmacol.*, **6**, 520.
56 Carson, S., Goldhammer, R., Weinberg, M. S., 1966, *Ann. N. Y. Acad. Sci.*, **130**, 935.
57 Schabort, J. C., 1967, *J. S. Afr. Chem. Inst.*, **20**, 103.
58 Strong, L. C., 1932, *Proc. Soc. Exp. Biol. Med.*, **30**, 386.
59 Strong, L. C., 1934, *Am. J. Cancer*, **20**, 387.
60 Strong, L. C., 1938, *Science*, **87**, 144.
61 Strong, L. C., Whitney, L. F., 1938, *Science*, **88**, 111.
62 Strong, L. C., 1939, *Am. J. Cancer*, **35**, 401.
63 Boyland, E., 1940, *Biochem. J.*, **34**, 1196.
64 Dittmar, C., 1940, *Z. Krebsforsch.*, **49**, 515.
65 Ciaranfi, E., 1972, *Arch. Sci. Med.*, **113**, 266.
66 Ciaranfi, E., Loreti, L., Borghetti, A., Guidotti, G. G., 1965, *Eur. J. Cancer*, **1**, 147.
67 Borghetti, A. F., Giglioni, B., Ottolenghi, S., Guidotti, G. G., 1970, *Biochem. J.*, **117**, 67P.
68 Ciaranfi, E., Perin, A., Sessa, A., Arnaboldi, A., Scalabrino, G., 1971, *Eur. J. Cancer*, **7**, 17.
69 Perin, A., Sessa, A., Scalabrino, G., Arnaboldi, A., Ciaranfi, E., 1972, *Eur. J. Cancer*, **8**, 111.
70 Mackenzie, C. G., Harris, J., 1957, *J. Biol. Chem.*, **227**, 393.
71 Vescia, A., Romano, M., Cerra, M., 1964, *Boll. Soc. Ital. Biol. Sper.*, **40**, 2047.
72 Weitzel, G., Schneider, F., Fretzdorff, A. M., Seyusche, K., Finger, H., 1964, *Krebsforsch. Krebsbekämpfung, Vol. V*, pp.156-162 [special number of *Strahlentherapie*, **57** (1964)].
73 Eaton, M. D., 1937, *J. Immunol.*, **33**, 419.
74 Eaton, M. D., 1938, *Bacterial Rev.*, **2**, 3.
75 Pappenheimer, A. M., 1938, *J. Biol. Chem.*, **125**, 201.
76 Arthus, M., 1930, *J. Physiol. Pathol. Gen.*, **28**, 529.
77 Boquet, P., Vendrely, R., 1943, *C. R. Soc. Biol.*, **137**, 179.
78 Heymans, M., 1926, *Arch. Int. Pharmacodyn. Ther.*, **32**, 101.
79 Együd, L. G., 1967, *Curr. Mod. Biol.*, **1**, 14.
80 Kulshrestha, D. C., Marth, E. H., 1970, *J. Milk Food Technol.*, **33**, 305.
81 Salfinger, M., 1970, *Pathol. Microbiol.*, **36**, 277.
82 Pepper, R. E., Chandler, V. L., 1963, *Appl. Microbiol.*, **11**, 384.
83 Woodland-Hastings, J., Gibson, Q. H., 1963, *J. Biol. Chem.*, **238**, 2537.
84 Woodland-Hastings, J., Spudich, J., Malnic, G., 1963, *J. Biol. Chem.*, **238**, 3100.
85 Fries, N., 1961, *Svensk Bot. Tidskr.*, **55**, 1.
86 Nyman, B., 1966, *Physiol. Plant.*, **19**, 377.
87 Norrman, J., 1968, *Arch. Microbiol.*, **61**, 128.
88 Van Buskirk, J. J., Frisell, W. R., 1967, *Biochem. Biophys. Acta*, **143**, 292.
89 Tyler, D. D., 1969, *Biochem. J.*, **111**, 665.
90 Rehak, M. J., Truitt, E. B., 1958, *Q. J. Stud. Alcohol*, **19**, 399.
91 Majchrowicz, E., 1965, *Can. J. Biochem. Physiol.*, **43**, 1041.
92 Kiessling, K. H., 1963, *Exp. Cel Res.*, **30**, 569.
93 Beer, C. T., Quastel, J. H., 1958, *Can. J. Biochem. Physiol.*, **36**, 531.
94 Hedlund, S. G., Kiessling, K. H., 1969, *Acta Pharmacol. Toxicol.*, **27**, 381.
95 Kiessling, K. H., 1963, *Acta Chem. Scand.*, **17**, 2113.
96 Weitzel, G., Buddecke, E., Schneider, F., 1961, *Hoppe-Seyler's Z. Physiol. Chem.*, **323**, 211.
97 Weitzel, G., Buddecke, E., Schneider, F., Pfeil, H., 1961, *Hoppe-Seyler's Z. Physiol. Chem.*, **325**, 65.

98 Weitzel, G., Schneider, F., Pfeil, H., Seynsche, K., 1963, *Hoppe-Seyler's Z. Physiol. Chem.*, **331**, 211.
99 Weitzel, G., Schneider, F., Fretzdorff, A. M., Seynsche, K., Finger, H., 1963, *Hoppe-Seyler's Z. Physiol. Chem.*, **334**, 1.
100 Weitzel, G., Schneider, F., Fretzdorff, A. M., 1964, *Experientia*, **20**, 38.
101 Weitzel, G., Schneider, F., Seynsche, K., Finger, H., 1964, *Hoppe-Seyler's Z. Physiol. Chem.*, **336**, 107.
102 Weitzel, G., Schneider, F., Fretzdorff, A. M., Seynsche, K., Finger, H., 1964, *Hoppe-Seyler's Z. Physiol. Chem.*, **336**, 271.
103 Weitzel, G., Schneider, F., Kummer, D., Ochs, H., 1968, *Z. Krebsforsch.*, **70**, 354.
104 Gutman, M., Mersmann, K., Luthy, J., Singer, T. P., 1970, *Biochemistry*, **9**, 2678.
105 Zollner, H., 1973, *Biochem. Pharmacol.*, **22**, 1171.
106 Haugaard, W., Lee, W. H., Kostrzena, R., Horn, R. S., Haugaard, E. S., 1969, *Biochim. Biophys. Acta*, **172**, 198.
107 Meijer, A. J., Groot, G. S. P., Tager, J. M., 1970, *FEBS Lett.*, **8**, 41.
108 Rosenkranz, H. S., 1972, *Bull. Environ. Contam. Toxicol.*, **8**, 242.
109 Hoard, D. E., 1960, *Biochim. Biophys. Acta*, **40**, 62.
110 Haselkorn, R., Doty, P., 1961, *J. Biol. Chem.*, **236**, 2738.
111 Sawicki, E., Sawicki, C. R., 1975, *Aldehydes—Photometric Analysis, Volume 2* (Academic Press, London) page 210.
112 Esplin, D. W., 1970, *The Pharmacological Basis of Therapeutics*, fourth edition, Eds L. S. Goodman, A. Gilman (Macmillan, London) page 1032.
113 Gleason, M. N., Gossolin, R. E., Hodge, H. C., Smith, R. P., 1969, *Clinical Toxicology of Commercial Products*, third edition (Williams and Wilkins, Baltimore).

α,β-Unsaturated aldehydes

3.1 2-Alkenals

3.1.1 *Structure of α,β-unsaturated aldehydes*

As a rule α,β-unsaturated aldehydes have the *trans* configuration. Since the resonance energy of the π-electrons is somewhat larger in the *trans*- than in the *cis*-isomers, the *trans* configuration is the more stable one; commercial crotonaldehyde, for example, contains only about 1% of the *cis*-isomer[1-3]. With certain synthetic procedures it has been possible to obtain 2-*cis*-alkenals[4], though during the formation of derivatives, such as 2,4-dinitrophenylhydrazones, the double bond rearranges to the *trans* configuration[4]. 2-Alkenals have a characteristic UV light absorption. The K bands ($\pi-\pi^*$) of the conjugated double bond system has an absorption maximum in the region 235–210 nm, with a molar extinction coefficient (ϵ) of 10000–15000. The R band ($n-\pi^*$) of the aldehyde group lies in the region 345–300 nm, $\epsilon = 20-100$. On changing the solvent from water to hexane, both the K band and the R band shift to shorter wavelengths. Only a few relevant studies on UV-spectroscopic data are cited here[5-10]. The aldehyde group of 2-alkenals is hydrated only to a slight extent in aqueous solution[11].

3.1.2 *Reactions with biologically important groups*

2-Alkenals are chemically highly reactive substances and they readily react even at neutral pH and at room temperature with biochemically important groups, such as sulphydryl, amino, or hydroxyl. Because of the conjugated double bond system several mesomeric forms exist, and nucleophilic reagents react either at C-3 in a 1,4-addition reaction, or at C-1 in a 1,2-addition reaction, or both at C-1 and C-3 [12]. Under physiological conditions the main reaction appears to be 1,4-addition.

3.1.2.1 Reactions with sulphydryl groups

The mechanism of reaction of α,β-unsaturated aldehydes with sulphydryl compounds and the structure of the reaction products have been investigated by many authors[13-20]. A comprehensive review on the reactions of thiols with various vinylogous carbonyls is given by Friedman[20a]. The results found in 1970 by Esterbauer[19] may be summarized as follows. One molecule of the sulphydryl compound reacts with one molecule of α,β-unsaturated aldehyde in a 1,4-addition reaction. The reaction product initially formed is a saturated aldehyde in which the sulphydryl group has been introduced at C-3 as a thioether linkage:

$$\text{RCH=CHCHO} + \text{R'SH} \longrightarrow \underset{\underset{\text{SR}'}{|}}{\text{RCHCH}_2\text{CHO}}$$

In the case of 4-hydroxy-2-alkenals the primary product cyclizes, with formation of a cyclic hemiacetal (cf section 3.2). The equilibrium lies well over to the side of the cyclic form, and no free aldehyde could be detected by IR, NMR, or UV spectroscopy. With 2,4-dinitrophenylhydrazine, however, a hydrazone derivative is formed, and with Schiff's reagent a positive aldehyde reaction is obtained. This shows that under certain conditions the cyclic form may rearrange back to the oxo form. Scheme 3.1 gives the most important reaction steps. The sulphydryl anion, formed in a prior reaction, reacts with the aldehyde to give a negatively charged resonant intermediate, which may either decompose again into aldehyde and sulphydryl ion, or react with a proton donor (HX) present in solution to give the end product. The latter, proceeding via the intermediate, may dissociate again into the initial reactants. Experiments show that the overall reaction is reversible and leads to an equilibrium. The formation of the end product proceeds with the kinetics of a second-order reaction, whereas the dissociation of the end product takes place with the kinetics of a first-order process.

$$RS^- + R'CH=CHCHO \underset{k_b}{\overset{k_a}{\rightleftharpoons}} \underset{\underset{SR}{|}}{R'CHCH\!\!\cdots\!\!CH\!\!\cdots\!\!O} \underset{(-HX)\, k_{X^-}}{\overset{(+HX)\, k_{HX}}{\rightleftharpoons}} \underset{\underset{SR}{|}}{R'CHCH_2CHO} + X^-$$

Scheme 3.1

It is generally assumed that the rate-determining step of the 1,4-addition reaction is the addition of the RS^- ion to the aldehyde, and that stabilization of the intermediate by means of hydroxonium ion (H_3O^+) or water occurs comparatively very rapidly. The kinetic analysis of a series of reactions revealed, however, that this is only the limiting case and that much more generally both reactions are rate-determining. In this case the rate of reaction (v) is given in equation 1 (scheme 3.2). The apparent rate constants, as determined experimentally, are given in equation 2 (scheme 3.2). The corresponding equations for the rate constants of the reverse reaction, and for the equilibrium constants, are readily derived. It follows from equation 2 that the reaction may be catalysed by hydroxyl ions (increase in α_{SH}) as well as by acids, HX (increase of the quotient).

Equation 1: $v = k_1 [\text{aldehyde}] [RSH + RS^-]$

{where v is the rate of formation of $RCH(SR)CH_2CHO$}

Equation 2: $k_1 = k_a \alpha_{SH} \dfrac{\sum k_{HX}[HX]}{k_b + \sum k_{HX}[HX]}$

{where α_{SH} is the degree of dissociation of RSH, and

$\sum k_{HX}[HX] = k_{H_2O}[H_2O] + k_{H_3O^+}[H_3O^+] + \ldots + k_{HX}[HX]$}

Scheme 3.2

In general it is found that hydroxyl ion catalysis predominates, so that with increasing pH the rate of the reaction increases or remains the same. The simplest conditions are then obtained if as HX component only the species H_3O^+ and H_2O may be involved, that is when the reaction occurs in buffer-free solution, and the pH values are adjusted according to the pH-stat principle with hydrochloric acid or sodium hydroxide. In figure 3.1 the results are shown of some measurements of this kind.

If the rate-determining step is the addition of the RS^- ion to the aldehyde, the reaction can be catalysed only by hydroxyl ions, and the kind and concentration of a buffer has no influence. The quotient of equation 2 (scheme 3.2) is equal to unity with these reactions. If the stabilization of the intermediate is also a rate-determining reaction step, then the velocity of reaction becomes greater, the greater the concentration of the species HX. As proton donors (HX compounds) those of greatest significance are buffer acids (CH_3COOH, $H_2PO_4^-$, HPO_4^{2-}, etc). The extent of the catalytic effects of buffers is very different for different reactions. Thus reactions of cysteine may be accelerated not at all, those of glutathione moderately, and those of thioglycolic acid esters very strongly (see table 3.1). For a specific reaction the extent of its acceleration is dependent on the pH value and the concentrations of the buffer (figure 3.2).

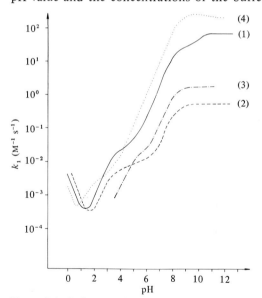

Figure 3.1. Influence of the hydrogen ion concentration on the rate constants k_1 of the reaction between α,β-unsaturated aldehydes and sulphydryl compounds:

$RCH=CHCHO + RSH \xrightarrow{k_1} RCH(SR)CH_2CHO$. The pH values were adjusted with hydrochloric acid or sodium hydroxide. (1) Crotonaldehyde + thioglycolic acid; (2) crotonaldehyde + thioglycolic acid ethyl ester; (3) crotonaldehyde + glutathione; (4) 4-hydroxypentenal + cysteine.

The reactivity of the α,β-unsaturated carbonyl system (C=C—C=O) towards sulphydryl compounds is greatly influenced by the substituents present. This is shown in table 3.2 with reference to equilibrium constants found for the reaction of some α,β-unsaturated carbonyl compounds with glutathione. The most reactive compound is acrolein, as expected; it reacts with the sulphydryl group not only the most rapidly of all carbonyl compounds, but also forms the most stable adduct. The reactivity of different 2-alkenals increases in the sequence: higher 2-alkenals < 4-hydroxy-2-alkenals ≪ acrolein. A measure of the stability of the adducts formed is given by their dissociation constants, and stability increases in the same order: higher 2-alkenals, 4-hydroxy-2-alkenals, acrolein. In comparing different α,β-unsaturated carbonyl systems, the following order of reactivity towards sulphydryl compounds, and of the

Table 3.1. Influence of 0·1 M phosphate buffer, pH 7·4, on the rate of reaction of α,β-unsaturated aldehydes with sulphydryl compounds.

Reaction	k_1 (M^{-1} s^{-1})		Acceleration of the reaction
	without phosphate	with phosphate	
Cysteine + crotonaldehyde	16·5	16·5	0
Cysteine + 4-hydroxypentenal	25·0	25·0	0
Glutathione + crotonaldehyde	0·25	1·25	5
Ethyl thioglycollate + crotonaldehyde	0·10	2·00	20
Ethyl thioglycollate + 4-hydroxypentenal	0·31	12·5	40

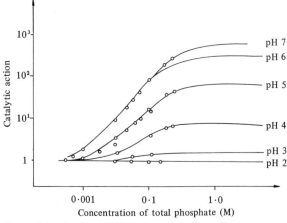

Figure 3.2. Catalytic effect of phosphate at different pH values on the reaction of 4-hydroxypentenal with ethyl thioglycolate.

Catalytic action = $\dfrac{\text{reaction velocity in solution containing phosphate}}{\text{reaction velocity in phosphate-free solution}}$.

stability of the adducts formed, generally applies: acid < ester < ketone < aldehyde (cf table 3.3). The constants given in table 3.2 provide a partial explanation for many very different biochemical and physiological effects of α,β-unsaturated aldehydes. From investigations with 4-hydroxy-alkenals it is known that many biochemical and physiological effects (toxicity; cytostatic action; inhibition of protein, DNA and RNA syntheses) have their origin in the blocking of essential sulphydryl groups (cf section 3.2). The dissociation constants given in table 3.2 are indeed valid for glutathione adducts, but one may assume that the dissociation constants of other adducts, in particular the reaction products of protein sulphydryl groups with aldehydes, would not be significantly different. The extent of the deactivation of sulphydryl groups depends on the aldehyde concentration and on the dissociation constant. In figure 3.3 this is presented graphically for several aldehydes. In the presence of a

Table 3.2. Equilibrium constants and reaction rate constants for the reaction of GSH and α,β-unsaturated carbonyl compounds in 0·067 M-phosphate buffer, pH 7·4, at 20°C.

$$\text{GSH} + \alpha,\beta\text{-unsaturated carbonyl} \underset{k_2}{\overset{k_1}{\rightleftarrows}} \text{adduct}.$$

k_1: second-order rate constant for adduct formation, in $M^{-1} s^{-1}$.
k_2: first-order rate constant for adduct dissociation, in s^{-1}.
$K = k_2/k_1$ is the equilibrium constant, which is the dissociation constant of the adduct (M^{-1}).

Carbonyl class	Substance	k_1	$10^7 k_2$	$10^7 K$
2-Alkenals	acrolein	121	17·5	0·14
	crotonaldehyde	0·785	301	416
	pentenal	0·471	344	738
	hexenal	0·330	470	1425[a]
	citral	0·0323	50	1630[a]
4-Hydroxy-alkenals	4-hydroxypentenal	2·19	4·31	2·96
	4-hydroxyhexenal	1·56	4·10	2·63[a]
	4-hydroxyheptenal	1·83	4·57	2·50[a]
	4-hydroxyoctenal	1·74	7·49	4·89
	4-hydroxynonenal	1·09	8·30	7·65[a]
	4-hydroxydecenal	1·96	9·50	4·85[a]
	4-hydroxyundecenal	1·47	9·36	6·38
	4-hydroxydodecenal	2·44	12·6	5·42[a]
2-Alkenone	methylvinyl ketone	31·9		
	methylisopropenyl ketone	0·60	77 000	127 000
	mesityl oxide	0·0023	112	48 000
	2-cyclohexen-1-one	0·336	51·6	153·5[a]

[a] These values were calculated from the corresponding k_2 and k_1 values; all other K values were experimentally determined.

Table 3.3. Toxicity, inhibition of DNA synthesis, and reactivity of α,β-unsaturated carbonyl compounds.

Carbonyl class	Substance	Toxicity (mM)		Inhibition of DNA synthesis [c]	Reactivity (k_1) [d]
		mice (LD_{50}) [a]	microorganisms [b]	10^{-8} mol per 10^6 EATC	(M^{-1} s^{-1})
2-Alkenals	acrolein	0.103	0.027	0.25	121
	methacrolein	—	0.050	—	—
	crotonaldehyde	2.3	0.20	6.0	0.78
	pentenal	2.9	—	17.5	0.47
	hexenal	3.0	—	8.75	0.33
	citral	2.9	—	—	0.032
2-Alkenone	methylvinyl ketone	1.1	0.022	2.5	31.9
	methylisopropenyl ketone	7.2	0.42	38.0	0.60
	mesityl oxide	>9.6	—	—	0.002
	cyclohexenone	1.8	—	26.0	0.34
Esters	methyl acrylate	—	0.68	—	—
	ethylcrotonate	>12.0	>2.6	—	0.003
Acids	acrylate	—	1.4	—	0.001
	crotonate	—	2.9	—	0.000
4-Hydroxyalkenals	4-hydroxypentenal	1.0	—	4.8	2.19
	4-hydroxyhexenal	1.0	—	8.1	1.56
	4-hydroxyheptenal	1.2	—	5.6	1.83
	4-hydroxyoctenal	1.46	—	4.6	1.74

[a] LD_{50} = single dose (mmol/kg^{-1} body weight) administered intraperitoneally to mice[39].
[b] Toxicity to microorganisms is the threshold concentration (mM) which inhibits bacterial growth[38].
[c] Inhibition of DNA synthesis: 50% inhibition of [^{3}H] thymidine incorporation into the acid-insoluble fraction of EATC[40].
[d] Reactivity: k_1 is the rate constant (M^{-1} s^{-1}) for the reaction of glutathione with the indicated carbonyl compounds (see table 3.2).

concentration of approximately 10^{-8} M free acrolein, 50% of the sulphydryl groups are deactivated. For this effect, a concentration of approximately 10^{-6} M of the hydroxyalkenal is required. Other 2-alkenals, such as crotonaldehyde, pentenal, or citral, influence sulphydryl groups only in concentrations of 10^{-3}–10^{-4} M.

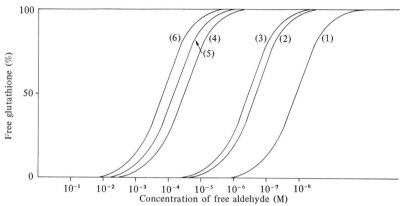

Figure 3.3. Free glutathione in the presence of different concentrations of aldehydes: (1) acrolein, (2) 4-hydroxypentenal, (3) 4-hydroxyoctenal, (4) crotonaldehyde, (5) pentenal, (6) citral.

$$\text{Free GSH} = \frac{100K}{K + \text{free aldehyde}},$$

where K is the dissociation constant of the GSH-aldehyde adduct (see table 3.2).

3.1.2.2 Reactions with amino groups

So far no detailed study has been made of the reaction of α,β-unsaturated aldehydes with amino groups of biologically important molecules. From the few available facts it would appear that 1,4-addition reactions as well as the formation of Schiff's bases are possible. The chemical constitution of the corresponding aldehyde, of the amino compound, and also the reaction conditions, all determine which compound is formed as the main reaction product (scheme 3.3).

```
RCHCH₂CHO    ←————————  RCH=CHCHO  ————→  RCH=CHCH=NR        (III)
    |                  (I)                        ↓
   NHR                                       RCHCH₂CH=NR
    ↓                                              |                (IV)
RCHCH₂CHO                                         NHR
    |                                              ↓
   NR          (II)                          RCHCH₂CH=NR
    |                                              |
RCHCH₂CHO                                         NR                (V)
                                                   |
                                             RCHCH₂CH=NR
```

Scheme 3.3

The Schiff's bases (III, IV, V) as well as the 1,4-addition products (I, II, IV, V) are unstable and readily split into the component aldehyde and amino compounds.

Cavins and Friedman[23] have shown that acrylonitrile, which undergoes 1,4-addition reactions in the same way as do α,β-unsaturated aldehydes, reacts with proteins at the ϵ-amino group of the lysine residue to give a mixture of mono- and di-substituted derivatives (I, II). Sulphydryl groups of amino acids or proteins react under comparable conditions some three hundred times as fast as amino groups[16], and the sulphydryl addition products are probably more stable by several orders of magnitude than the amino addition products. In a biological system which contains sulphydryl as well as amino compounds, α,β-unsaturated aldehydes will therefore react almost exclusively with the sulphydryl groups, and only under special conditions are amino groups also attacked. Acrolein is the most reactive aldehyde, and it has been shown that this aldehyde in high concentrations is an excellent reagent for the introduction of cross-linkages into proteins: collagen[24-30], elastin[31], and connective tissue[29,32] have been particularly examined in this respect. Bowes and Cater[27] have shown that in collagen treated with acrolein the ϵ-amino groups of lysine have been modified. 1,4-Addition reactions, Schiff's base formation, and presumably also aldol condensations, participate in the development of cross-linkages (scheme 3.4). On the basis of its cross-linking action acrolein has been recommended as a fixative for many biochemical investigations[33,34,35,36]. Crotonaldehyde and higher alkenals are appreciably less effective and produce hardly any cross-linking in proteins. Alarcon[37] has reported that acrolein used in low concentrations does not react with the amino groups of naturally occurring amines at neutral pH.

protein—$SCH_2C=CHCH_2CH_2S$—protein
  |
  CHO

protein—$NHCH_2C=CHCH_2CH_2NH$—protein
  |
  CHO

protein—$SCH_2CH_2CH=N$—protein

protein—$NHCH_2CH_2CH=N$—protein

Scheme 3.4

3.1.3 *Toxicity of α,β-unsaturated aldehydes*
3.1.3.1 Relation between structure and toxicity
Stack[38] has examined the toxicity of many α,β-unsaturated carbonyl compounds towards microorganisms, and found that they can inhibit biological oxidations. The toxicity of analogous compounds rapidly decreases in the sequence: aldehyde, ketone, ester, and acid (table 3.3).

Of the α,β-unsaturated aldehydes and ketones the most poisonous are respectively acrolein and methyl vinyl ketone. If the hydrogen atoms in the vinyl group of the compound $CH_2=CHCHO$ are replaced by alkyl residues, the toxicity rapidly decreases. In table 3.3 comparative toxicity values found for mice on intraperitoneal injection[39] are given. It is noteworthy that the same sequence applies here as with bacteria.

Naturally the limits of toxicity for every member of the above series vary for different organisms, but differences in toxicity between the members of the series remain substantially the same. It is interesting to note that inhibition of DNA synthesis parallels the toxicity[40,41].

The toxic effect is ascribed to the presence of α,β-unsaturated carbonyl grouping. The corresponding saturated aldehydes (for example, propionaldehyde) or α,β-unsaturated alcohols (for example, allyl alcohol) are appreciably less poisonous[38,42]: as mentioned in section 3.1.3.2, allyl alcohol can be metabolized in liver to acrolein[61].

In the previous section it was shown that α,β-unsaturated carbonyl compounds react under physiological conditions with sulphydryl groups mainly by 1,4-additions. Sulphydryl groups in proteins and in lower molecular weight compounds such as glutathione play an important role in the living cell. The deactivation of these essential sulphydryl groups leads to interference with intermediary metabolism, and to an inhibition of cell growth and of cell division. A comparison of toxicity values and reaction rate constants shows that the more toxic an α,β-unsaturated carbonyl compound is, the faster it reacts with sulphydryl groups (table 3.3). This points to a competition between the blocking of functional sulphydryl groups on the one hand, and the detoxification of the substance on the other—a substance which reacts very slowly with sulphydryl groups is detoxified before it can exert its toxic effects. For α,β-unsaturated aldehydes several detoxification mechanisms may be possible, namely the enzyme-catalysed addition to glutathione, the enzyme-catalysed oxidation to unsaturated acids, or the reduction to alcohols (cf section 3.3). When given sublethal concentrations of a carbonyl compound, microorganisms develop a resistance towards 2-alkenals but not towards α,β-unsaturated ketones. Acrolein being the most reactive unsaturated aldehyde, is thus appreciably more toxic than are other α,β-unsaturated aldehydes towards the mammalian organism and towards microorganisms.

3.1.3.2 Toxic effects of acrolein

There is no doubt that acrolein is one of the toxic components of atmospheric 'smog'. The sources of the acrolein content of air are automobile exhaust gases, as well as waste gases from factories which produce synthetics and manufacture fats[54,119-121]. Numerous experiments have been carried out with human beings, animals, and plants, to determine what physiological and biochemical changes are caused by acrolein and what concentrations of acrolein are tolerable. These investigations are the more significant, since it is known that acrolein occurs in low concentration also in cigarette smoke[122], so that for smokers there is the hazard of being exposed to additional quantities of acrolein.

After only a short time, acrolein vapour (0·80 ppm) causes painful irritation of the eyes and of the external mucous membranes. On prolonged exposure to acrolein pathological changes occur, mainly in the respiratory organs[43,44]. A condition resembling asthma sets in and is accompanied by progressive paralysis of respiration and of circulation[45]. In the lung there is inhibition of the lactate dehydrogenase[46]. Inhaled, or intra-peritoneally injected, acrolein causes a very marked increase in liver alkaline phosphatase[47] and tyrosine transaminase[48] activities. In the blood it produces a decrease of cholinesterase activity[49,50] and a decrease of the polymorphonuclear leucocytes[49]. Murphy and Porter[51] found that after intraperitoneal injection of small amounts of acrolein (1·5 mg kg^{-1}) into rats glycogen storage is increased. This is caused by the secretion, stimulated by acrolein, of glucocorticoids which in their turn stimulate the synthesis of adaptive liver enzymes concerned in gluconeogenesis. In erythrocytes glucose uptake is inhibited; glucose degradation by the Embden–Meyerhof pathway is decreased, and the phosphogluconate pathway is inhibited, thereby decreasing the supply of NADPH[52]. In partially hepatectomized rats the injection of acrolein (1 mg kg^{-1} body weight) causes an inhibition of DNA and RNA synthesis in liver and lung[113]. Experiments *in vitro* showed that DNA polymerase isolated from regenerating liver is inhibited by acrolein, whereas DNA polymerase from *Escherichia coli* is activated[114]. In both cases the sulphydryl group of the enzyme is modified; in the first case the sulphydryl residue is essential for the enzymatic activity, but in the second case it is not[115,116]. A series of other isolated enzymes with functional sylphydryl groups, such as lactate dehydrogenase, alcohol dehydrogenase, and glucose 6-phosphate dehydrogenase[117], as well as *L*-asparaginase from *Escherichia coli*[118], are deactivated by acrolein.

In the USSR the maximal permissible concentration of acrolein in the atmosphere is 0·043 ppm (0·11 mg m^{-3})[49,53]. More recent long term experiments showed, however, that even this concentration still produced statistically significant pathological changes in rats, so that a maximal permissible concentration of approximately 0·013 ppm (0·03 mg m^{-3}) was recommended[54]. In drinking water the concentration must not exceed 0·01 mg l^{-1} [55]. The raised respiratory flow resistance during acrolein inhalation is caused by cholinergic nerve stimulation, which acts on the bronchial muscles[56]. It is also of interest that acrolein and the higher 2-alkenals have a digitalis-like action, and can similarly inhibit the ion transport–adenosine triphosphatase system in heart muscle[57]. The active group of digitalis compounds is known to be the carbonyl group conjugated with an ethylenic double bond; the remainder of the molecule contains a group which facilitates a specific binding to the receptor site. 2-Alkenals have only the active grouping and are therefore much less specific in their cardiovascular actions. Izard[58] and Puiseux-Dao and Izard[59] tested the toxicity of aldehydes, which are present in cigarette smoke (acrolein, crotonaldehyde, formaldehyde, acetaldehyde, propionaldehyde), on algae (*Dunaliella bioculata*), and found the highest toxicity with acrolein. Even in concentrations of 0·14 mM, acrolein can induce a change in the nuclear structure, can block cell division, and can induce mutations. On the basis of its inhibitory effect on plant growth acrolein has been recommended for the destruction of weeds in stagnant or slowly flowing waters[60].

Rees and Tarlow[61] have reported that the liver-damaging activity of allyl formate is due to enzymatically produced acrolein. In the periportal zone of the liver lobule allyl alcohol, formed from allyl formate, is oxidized to acrolein by nonspecific liver alcohol dehydrogenases, which may then cause histochemically observable necrosis in that location.

In table 3.4 some toxic effects of inhaled acrolein are summarized. Further details of the toxic effects of acrolein are given by Champeix and Pierre[63]. Crotonaldehyde engenders similar pathological changes to those caused by acrolein, but only at higher concentrations. The toxic effects of crotonaldehyde are reported by Skog[64], Rinehart[65], Trovimov[66], and Sim and Pattle[43].

Table 3.4. Toxic effects of inhaled acrolein. 1 ppm = 2·35 mg m^{-3}.

Concentration	Effects, etc.
0·013–0·043	maximum permitted concentration in the atmosphere[43,49,55]
0·35	limit of concentration for detection by smell[43]
0·22–0·75	irritation of the eyes[43] difficulty in breathing[43,44] alteration of the reflexes[49] increase of luminescent leucocytes[49] decrease of cholinesterase activity[49,50]
0·4–1·0	lachrymation and painful irritation of the mucous membranes after a few seconds[43] increase of the respiratory flow resistance and prolonged deep respiration[56] inhibition of lactate dehydrogenase in the lung[46]
2·1	alteration of the activity of liver enzymes[47]
50	acutely toxic to mice[62]

3.1.4 Antitumour effects of α,β-unsaturated aldehydes

Especially detailed investigations have been carried out on citral (*cis* and *trans*-2,6-dimethyl-2,6-octadien-8-al) (table 3.5). Citral belongs to the alicyclic aldehydes of the terpene series; it is the main component (approximately 80%) of lemon grass oil, and is found in all citrus fruits.

Its anticancer activity was first studied by Boyland and Mawson in 1938[67, 70], and by Dittmar in 1940[71], and these authors showed that citral retards the growth of spontaneous tumours and of inoculated tumours in mice. Acrolein and crotonaldehyde, which were also tested by Boyland, are active only against inoculated tumours. The corresponding acids, namely geranic acid, acrylic acid, and crotonic acid, are inactive. This shows that the α,β-unsaturated aldehyde grouping is responsible not only for the general toxic activity, but also for the carcinostatic action—an observation confirmed by Osato et al.[73]. Citral is also quite active towards tumours in man, and Osato[77] surveys twenty years of experience in tumour chemotherapy with citral. In these treatments the patients received daily intramuscular injections of approximately 5 ml of a 5% emulsion of citral over several weeks to several months. Of a total of one hundred and twenty-one patients with various advanced inoperable carcinoma, in six cases (five stomach carcinomas and one pulmonary metastasis) the tumour disappeared and the patients were completely cured. Carcinoma of the stomach seems to respond best to the treatment with citral; of eighty-two patients with advanced stomach cancer five were permanently healed (followed up from ten to fifteen years) and nine patients were much improved for at least some time (tumour disappeared and X-ray findings greatly improved). In nineteen other cases appreciable improvement occurred, that is to say patients showed general signs of recovery, tumours diminished in size, and X-ray findings were improved.

Of the other tumours treated (uterus, pancreas, liver, or lung carcinoma) only lung cancer responds significantly to citral treatment, three of fourteen such cases showing regression of the tumour. Morrow[75] reported on the treatment of advanced carcinoma of the bladder: of twenty-six patients who had been treated with citral for over four months, six showed

Table 3.5. Summary of types of tumours treated with 2-alkenals.

Compound	Tumour treated
Acrolein	inoculated tumours, mouse (sarcoma M.C.D.B.I.)[67]
	experimental leukaemia, mouse (L 14 A 4 R)[68,69]
Crotonaldehyde	inoculated tumours, mouse (sarcoma M.C.D.B.I.)[67]
Citral	spontaneous breast tumours, mouse[70]
	Ehrlich carcinoma solid, mouse[71]
	methylcholanthrene-induced tumours and inoculated tumours, mouse (M.C.D.B.I.)[67]
	stomach, colon, lung cancers, man[72]
	Yoshida sarcoma, mouse[73]
	stomach, colon, lung cancers, man[74]
	bladder carcinoma, man[75]
	adenocarcinoma, mouse[76]
	various carcinomata, man[77]
	HeLa cells[78]

a definite improvement. The regression, and the inhibition of the growth, of tumours may possibly be connected with capillary damage caused by the citral; the newly formed capillary system of the tumours may be particularly sensitive[71] to cytotoxic agents. Leach and Lloyd[79] found that citral, even in very low concentrations, causes damage to the endothelial tissue, though simultaneous application of vitamin A prevents damage to the vessels. According to Morrow[75] and Herzmann[80] citral can competitively displace vitamin A, which is necessary for metabolism in the vessels, and can thus act as an antivitamin A. For structural reasons it is likely that we are dealing with competition with vitamin A aldehyde.

Of particular interest is the work of Alarcon and Meienhofer[81] on the formation of acrolein during degradation *in vitro* of cyclophosphamide. Cyclophosphamide, a commonly used antitumour agent introduced by Arnold and Bourseaux[82], is itself inactive. Activation into a cytostatically effective form (scheme 3.5) takes place preferentially in the liver by microsomal hydroxylation (II). According to Brock and Hohorst[123] and Hill et al.[124] 4-hydroxycyclophosphamide (so-called 'aldophosphamide') is, either in the cyclized form (II) or as the free aldehyde (III), the active principle of cyclophosphamide. Alarcon and Meienhofer[81] first pointed out that the aldehyde (III) should be unstable and undergo a β-elimination to form acrolein. In fact Alarcon was able to show that liver microsomes can convert cyclophosphamide into acrolein *in vitro*, and this at a rate equal to the rate of activation of cyclophosphamide *in vivo*.

(I) $(ClCH_2CH_2)_2N-P$ (cyclic structure with O, H, N, O)

(II) $(ClCH_2CH_2)_2N-P$ (cyclic structure with O, H, N, OH, H)

(III) $(ClCH_2CH_2)_2N-P(=O)(NH_2)-O-CH_2-CH_2-CHO$

(IV) $(ClCH_2CH_2)_2N-P(=O)(NH_2)-OH + CH_2=CH-CHO$

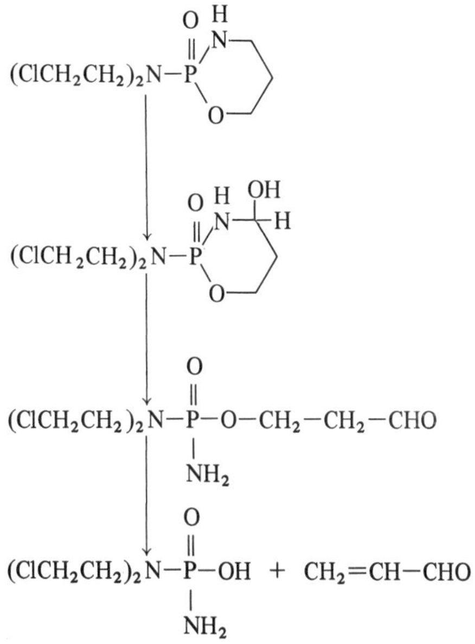

Scheme 3.5

It was suggested that the active principle of cyclophosphamide is acrolein, an aldehyde known to be highly cytotoxic from its ability to react with biological sulphydryl and amino groups. Later it was reported that acrolein is also formed by the degradation *in vitro* of isophosphamide, an isomer of cyclophosphamide[125]. Newer reports by Phillips[126] and Connors et al.[127] show that not only acrolein but also the other breakdown product phosphorodiamide-mustard (IV) is a highly cytotoxic compound and may thus be responsible, at least in part, for the antitumour activity of cyclophosphamide.

Although the formation of acrolein *in vitro* is not yet proved it is very probable that its formation actually takes place in the tumour cells. In any case it must be assumed that acrolein is formed directly in the target cell as it cannot be transported by the bloodstream, where it is immediately deactivated by reaction with glutathione and other sulphydryl compounds. Therefore acrolein itself cannot be used as a systemic antitumour agent. This is confirmed by Motycka and Iacko[69], according to whom leukaemic leucocytes are deactivated only by direct contact with acrolein.

3.1.5 Bactericidal and fungicidal effects of α,β-unsaturated aldehydes

In 1952 Hirsch and Dubos[84] showed that spermine shows bactericidal activity towards several mycobacteria in the presence of bovine serum. Further investigations[85] revealed that bovine serum contains an amine oxidase which oxidizes spermidine (I) and spermine (II) to the corresponding aminoaldehydes (III, IV) (scheme 3.6). These aminoaldehydes are toxic to bacteria, bacteriophages, plant and animal pathogenic viruses, and animal cells; for references see Kimes and Morris[86]. The dialdehyde from spermine is a more active inhibitor than the monoaldehyde derived from spermidine. Reduction of the aldehyde to the alcohol group destroys the

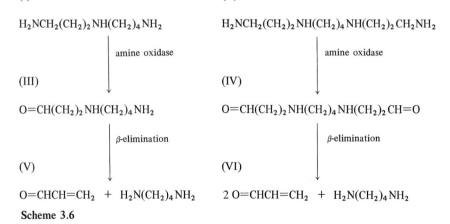

Scheme 3.6

bactericidal activity[87]. Aminoaldehydes (III) and (IV) are unstable compounds and spontaneously form acrolein (V) by β-elimination[88-90]. Alarcon[37,88], and Kimes and Morris[89], suppose that the acrolein formed *in situ* is chiefly responsible for the toxic effects of oxidized spermine and spermidine.

Acrolein, as has been shown in section 3.1.2, can readily modify proteins with formation of intramolecular cross-linkages, and can block sulphydryl-dependent cellular functions on the basis of its reactivity and affinity towards sulphydryl groups. Kimes and Morris[86] have shown that when acrolein acts on *E. coli* the DNA synthesis is the most susceptible process. With a concentration of acrolein as low as 9 µM the DNA synthesis is completely inhibited. With somewhat higher concentrations (13–20 µM) RNA synthesis and protein synthesis are also completely inhibited. Many other α,β-unsaturated aldehydes also inhibit DNA (cf table 3.3), and the growth limiting action of 2-alkenals probably proceeds in general by inhibition of the synthesis of macromolecules. In assessing the activity of substance (IV) (scheme 3.6), besides the formation of acrolein, one should consider the presence of two aldehyde groups, since it is well known that dialdehydes (such as glyoxal, malonaldehyde, or glutaraldehyde) possess strongly marked bactericidal, fungicidal, and virucidal properties. Stack[38] has examined the antimicrobial properties of a range of α,β-unsaturated aldehydes, and found that their toxicity depends on structure. The most active aldehydes are acrolein and crotonaldehyde, whose bactericidal action had been described by earlier authors[90,91]. It is of technical interest that acrolein is added in concentrations of 0·0001% to petroleum fractions, in order to stop the growth in oil tanks of microorganisms such as *Bacillus subtilis* or *Aspergillus niger*[92]. Citral likewise shows a strongly marked bactericidal activity associated with low toxicity towards mammals, and Osnos[93] recommends citral for the prophylaxis and therapy of postnatal infections. Citral, and other terpene aldehydes present in grasses, inhibit the growth of microbes in the stomachs of ruminants, and this may prejudice full utilization of their foodstuffs[94].

Geiger and Conn[95] examined the fungistatic action of several (mainly aromatic) compounds possessing the α,β-unsaturated carbonyl grouping (C=C—C=O). It was suggested that the fungistatic activity of these compounds is associated with their reactivity towards sulphydryl groups. An unequivocal connection between fungicidal action and reactivity of carbon–carbon double bonds towards nucleophilic reagents was found by McGowan *et al.*[96] for eighty different compounds; most of these contained the α,β-unsaturated carbonyl system. The most reactive compounds were those which had an electron-attracting group (CHO, COR, CO_2R, CO_2H) substituted at one of the carbon atoms of the double bond. One of the strongest electron-attracting groups is the nitro group, and nitroethylene derivatives do in fact show a very strong fungistatic activity[97].

2-Hexenal, present in the leaves of many plants, appears to have an important function in warding off microbial infections. As Schildknecht and Rauch[98] have shown, 2-hexenal is of low toxicity towards mammals, but shows marked fungicidal and bactericidal activity, and may also kill ciliates such as the slipper animalcule (*Paramecium caudatum*).

It has been supposed that 2-hexenal is the active principle of 'aerial phytoncide'. According to Tokin[99] this involves protistocidal, fungicidal, and bactericidal substances which are exuded specially by higher plants, thus surrounding the plant leaf and protecting it against microbial infections. According to Schildknecht and Rauch[98], hexenal is the phytoncide of robinia, oak, alder, lupin, blackcurrant, bilberry, cranberry, privet, and grasses. These authors suppose that hexenal is formed also by the undamaged leaf. In a twelve-hour field experiment with robinia, hexenal could in fact be shown to be present in the air within plastic bags which surrounded the plant. Other authors, on the other hand, tend to the opinion that hexenal is formed only after damage to the leaf. Major *et al.*[100] ascribed the well known resistance of the gingko tree (*Gingko biloba*) towards fungi to the production of hexenal as a response to slight damage of the leaf when a fungal infection starts.

A series of thiosemicarbazones of saturated and unsaturated aldehydes were tested by Manowitz and Walter[101] for their activity *in vitro* against different bacteria, yeasts, and fungi. The thiosemicarbazones of α,β-unsaturated aldehydes were found to be appreciably stronger antimicrobially than those of the saturated aldehydes. The effectiveness further depends on the chain-length, the most active being the unsaturated aldehydes of ten to twelve carbon atoms. Gingras *et al.*[102] believed that the fungicidal activity of the thiosemicarbazones depends on their ability to form complexes with copper ions; they reported that the greatest fungicidal activity was shown by the thiosemicarbazone of 2-undecanal.

Isonicotinic acid hydrazide[103,104] and a pyridylcarbazone of crotonaldehyde[105] are active against tubercle bacilli.

3.1.6 *Antiviral effects of α,β-unsaturated aldehydes*

Spermine aldehyde and spermidine aldehyde not only deactivate bacteria (cf section 3.1.5), but are also active against bacteriophages, plant viruses, and animal pathogenic viruses such as influenza virus, Newcastle disease virus, myxovirus, and vesicular stomatitis virus (for literature see Bachrach and Rosenkovitch[106]). Here, too, one should consider acrolein formed endogenously as the active principle. Thus, for example, 0·4 mM acrolein reduces the infectious titre of vaccinia virus by approximately three orders of magnitude within five hours. Acrolein therefore is equally as active as oxidized spermine, which reduces the titre by three and a half orders of magnitude within five hours at a concentration of 0·41 mM [106]. Nevertheless, Bachrach and Rosenkovitch[106], and Kremzner and Harter[107], believe that the antiviral activity of spermine and spermidine aldehydes may not be explained solely on the basis of acrolein formation. Kremzner and Harter[107] have compared the antiviral activities of spermine aldehyde and spermidine aldehyde with the activity of synthetic aldehydes (table 3.6). The results appear to indicate that the amino group in the aldehyde molecule is in fact not essential for the antiviral activity, since α,β-unsaturated aldehydes (acrolein, hexenal, citral) as well as some dialdehydes (glutaraldehyde, malonaldehyde) are equally as effective as are oxidized spermine or spermidine. Saturated monoaldehydes, but excluding octanal, show no antiviral activity. Spermine dialdehyde probably acts as a dialdehyde as well as a source of acrolein, but the monoaldehyde of spermidine acts only via acrolein formed *in situ*.

Table 3.6. Effect of aldehydes on the infectivity of vesicular stomatitis virus (VSV) according to Kremzner and Harter[107]. The virus was incubated at 37°C for 3 h with 0·4 mM aldehyde; the excess of aldehyde was removed by washing, and the infectious units of VSV were determined by the plaque technique.

Aldehyde	lg VSV deactivated (virus units)	Aldehyde	lg VSV deactivated (virus units)
Formaldehyde	0·4	Acrolein	1·6
Acetaldehyde	0	2-Hexenal	1·7
Propionaldehyde	0	Citral	3·7
Butanal	0·2	Malonaldehyde	3·8
Hexanal	0·2	Glutaraldehyde	5·2
Heptanal	0·1	*ortho*-Phthalaldehyde	5·1
Octanal	1–4	Spermidine aldehyde	3–5
		Spermine dialdehyde	3–5

Of the dialdehydes it is known that they may react directly *in vitro* and *in vivo* with the amino group of the bases in DNA and RNA, and that they may also modify the amino groups in the lysine residues of virus proteins (cf section 5). Spermine dialdehyde can also be irreversibly bound to the amino groups of DNA bases[108].

It is not known if acrolein and other α,β-unsaturated aldehydes react directly with RNA or DNA. It appears certain, however, that they are able to modify the protein part of viruses by formation of cross-linkages and by modification of the amino and the sulphydryl groups. It has been observed that acrolein and crotonaldehyde modify the chromosomes of *Drosophila* and produce mutations[109].

According to Tiffany *et al.*[110], crotonaldehyde is inactive against animal pathogenic viruses (Newcastle disease virus, influenza virus). However, crotonaldehyde appears to be very active against plant viruses and has been recommended for the treatment of virally infected plants[111,112].

3.2 4-Hydroxy-2-alkenals
3.2.1 *Occurrence in autoxidized polyene fatty acids*
The development of our own investigations, which are described in this section, shows several parallels to the work of Schubert (section 5.4). In both studies, on physical or chemical contact in aqueous solution with nutrient substrates, new substances were formed whose strongly marked cytotoxic activities are of biological interest. Schubert concerned himself with the anaerobic action of high energy radiation on unbuffered aqueous sugar solutions. Our work deals with the action of atmospheric oxygen on unbuffered aqueous emulsions of higher unsaturated fatty acids and fatty acid esters.

In both cases the substances finally investigated arise as products of secondary transformations of primary reaction products, for example a dicarbonyl sugar (Schubert). In each case hydrogen peroxide is formed as by-product and, in each case again, radical reactions play a decisive role.

Even the initial questions posed were very similar. Schubert was concerned with the problem of the biological activity of irradiated foodstuffs. In our investigations we wished to discover the biological activities of the higher unsaturated fats which had undergone autoxidation under physiological conditions. Such reactions take place when higher unsaturated fats are emulsified at 40°C in a large excess of distilled water, with exposure to the air[128]. The observation that water-soluble substances could be formed[129] was of primary interest to us. After excluding the possibility that these water-soluble products were impurities, leached out of the starting material[131], the first UV- and IR-spectrometric investigations were carried out[132]. These revealed in the water-soluble reaction products the presence of hydroxyl and carbonyl groups, a small but very constant proportion of conjugated unsaturated systems, as well as of aliphatic carbon chains. The original supposition, that the starting material would add water molecules to the double bonds, was discarded after it was shown that on rigorous exclusion of oxygen no reaction took place[133]. Active oxygen was found to be present in the aqueous phase some time after the aerobic dispersion of the starting material[134]. This result permitted the conclusion that the so-called 'water reaction' (Wasserreaktion) is essentially an oxidative alteration of the starting material, albeit under the special conditions of an aqueous emulsion.

After standardization of the experimental conditions (very pure methyl or ethyl linoleate, emulsified in fifty volumes of twice-distilled water) the total system was thoroughly investigated. The dispersed linoleate ester and the aqueous phase, but especially the reaction products dissolved in the latter, were analysed[135-140]. From all the analytical data and the chemical constitution of the reaction products so far isolated it has become certain and unambiguous that, in pure water, aerobically dispersed fatty acid esters undergo the usual autoxidation, and that the primary products are the expected monohydroperoxides. These latter are known to be highly reactive substances, and to react further to give carbonyl compounds via homolytic and heterolytic cleavage reactions.

From the example of methyl 9,12-linoleate it could be shown that the initially formed linoleic acid ester monohydroperoxides appear as unstable intermediate products, which rise to a maximum amount after thirty hours, and are converted into secondary and tertiary products with simultaneous formation of hydrogen peroxide. These consecutive reactions are specifically chain cleavage reactions, which produce oxygen-containing fragments of the original linoleic ester monohydroperoxide molecule. These fragments are now water soluble, and therefore transfer to the aqueous phase and may thence be isolated. Acidic and neutral fragments as well as hydrogen peroxide are present.

An initial investigation of the biological action of the total water-soluble products produced evidence of an undoubted effect on the oxidative metabolism of normal rat liver slices. This involved a 20%-50% inhibition of oxygen consumption, a 20%-30% inhibition of aerobic glycogenolysis, and an acceleration of the peroxidation of tissue lipids by 200%-300%[141]. In addition an unambiguous inhibition (up to 80%) of the alcoholic fermentation of yeast was demonstrated[142].

After this, the biological action of the three main components of the water-soluble reaction products (acid and neutral components, and hydrogen peroxide) were differentiated from one another[143].

It was found that the acid fraction exerts a marked inhibition on the respiration of healthy rat liver, on the anaerobic glycolysis and the respiration of yeast cells, as well as on the germination of cress seeds. The acid fraction, on the other hand, showed little inhibition of glycolysis and respiration of Ehrlich ascites tumour cells (EATC) in the mouse.

The neutral fraction proved strongly inhibiting in this test system of special biological interest. No more precise conclusions could be drawn about the active agent(s), because the constitution of the neutral fraction was unknown.

The results with EATC encouraged us to carry on the investigations initially with the neutral fraction. After identification and separation of the hydrogen peroxide, whose cytotoxic properties are known, there remained a mixture of substances, free from carboxylic acids, which gave a remarkably constant content of lipid hydroperoxides, and was thus termed preparation LHPO[144]. The biological inhibitions determined for LHPO were as follows: inhibition of anaerobic glycolysis and respiration in EATC[144]; lowering of the cytoplasmic NAD^+ concentration in EATC[144]; inhibition of glycolysis in yeast cells[144]; inhibition of aerobic glycolysis in EATC[145]; inhibition up to complete prevention of the development of EATC in the living animal after treatment of EATC with LHPO *in vitro*[146]; deactivation of the glyceraldehyde phosphate dehydrogenase (GAPDH) and of the lactate dehydrogenase (LDH) in EATC[147], and morphological and other cytotoxic damage to EATC[148].

The multifarious activity spectrum of this preparation, whose constitution was not yet known, prompted us to make an attempt to separate preparation LHPO chromatographically, with the aim of isolating and identifying the most important active agents as chemically pure substances.

It was demonstrated initially that a substantial part of the biological inhibitory activity disappeared when the active oxygen of the lipid hydroperoxides present had been destroyed by reduction[144-146]. The action of the hydroperoxide portion was explained by an oxidation of functional sulphydryl groups. The significant residual inhibitory activity present after reduction of the hydroperoxide clearly demonstrated the presence of further nonperoxidic inhibitors in preparation LHPO. New chromatographic methods of separation were developed, involving a combination of thin-layer and preparative column chromatography, and thus proved possible to isolate, and to identify chemically in an unambiguous manner, the pure substances from preparation LHPO[149], which are listed in table 3.7. As seen from the quantities given, they represent barely half of the total quantity of preparation LHPO. The remaining components, among which is a fourth peroxidic component, have not yet been identified.

Table 3.7. Survey of substances isolated so far from the water-soluble neutral components of autoxidized linoleic acid ester.

Number	Compound	g/100 g prepn LHPO	g/100 g linoleic acid ester	Biological inhibitory activity	Reference
(I)	2-Octen-1-al	2·0	0·07	not investigated	137
(II)	1-Hydroperoxy-pentane	2·0	0·07	+	137, 149
(III)	8-Hydroperoxy-caprylic methyl ester	9·0	0·315	+	138, 150
(IV)	2-Oxo-heptan-1-ol	2·0	0·07	+	139
(V)	4-Hydroperoxy-2,3-nonen-1-al	2·5	0·088	+	138, 150
(VI)	4-Hydroxy-2,3-octen-1-al	3·8	0·133	+	139, 148, 151–159
(VII)	Hydroxycaprylic acid methyl ester	6·5	0·227	not investigated	129, 152
(VIII)	ω-Carbomethoxy-2-oxo-alkan-1-ol	7·5	0·263	not investigated	129
(IX)	ω-Carbomethoxy-4-hydroxy-2,3-alken-1-al	7·5	0·263	not investigated	129

Nevertheless the three hydroperoxides identified (II, III, and V) can explain the major inhibitory activities of the peroxide components of LHPO. The inhibitory effects are those on respiration, on aerobic and anaerobic glycolysis, as well as on the activities of GAPDH and LDH in EATC[149,150].

From the standpoint of this book the carbonyl compounds isolated from LHPO (substances I, IV, V, VI, VIII, and IX) are of particular interest. Whereas substance (I) is an aldehyde of the well-known type, the others are carbonyl compounds so far unidentified. Substance (V), in particular, should be of chemical and biochemical interest, since it is a molecule which incorporates both the reducing aldehyde group and the strongly oxidizing hydroperoxide function. In consequence this compound is extremely labile and was particularly difficult to isolate. It is bifunctional not only in the chemical sense, but also as a metabolic inhibitor. One inhibitory function is clearly based on the hydroperoxy residue, which, as already mentioned, can oxidatively block functional sulphydryl groups. The second inhibition is associated with the α,β-unsaturated carbonyl system, which reacts with such sulphydryl groups and thereby deactivates them (see section 3.1). This double function explains why substance (V) showed the strongest activity in the inhibition test mentioned[150], as can be seen from the summary presented in table 3.8. Undoubtedly compound (V) would have been the most interesting substance for detailed biochemical and biological investigation, but this was not feasible because of the very considerable difficulty of isolation of this labile substance from LHPO, as well as the enormous difficulties of a possible chemical synthesis. Table 3.8 shows in addition the inhibitory activities of aliphatic α,β-unsaturated seven carbon atom and eight carbon atom aldehydes, and thus provides data on the activity of the second inhibitory grouping in substance (V). This system exerts its inhibitory effect as the result of a 1,4-addition of sulphydryl groups to the double bond between C-2 and C-3 (see also section 3.2.5). As table 3.8 shows, the activity of the α,β-unsaturated

Table 3.8. Inhibitory activities of several substances isolated from a preparation of LHPO, and of 2-alkenals, on the respiration and glycolysis of Ehrlich ascites tumour cells.

Substance	Glycolysis	Respiration
(V)	$1 \cdot 6 \times 10^{-4}$ M	$2 \cdot 7 \times 10^{-4}$ M
(III)	$3 \cdot 2 \times 10^{-4}$ M	$2 \cdot 0 \times 10^{-3}$ M
(II)	$3 \cdot 0 \times 10^{-4}$ M	$1 \cdot 5 \times 10^{-3}$ M
Other peroxides	$0 \cdot 5 - 1 \cdot 0 \times 10^{-3}$ M	
Seven carbon atom, or eight carbon atom 2,3-alkenals	$3 \cdot 0 \times 10^{-3}$ M	maximally only 40% inhibition at 4×10^{-3} M
(VI)	$3 \cdot 0 \times 10^{-4}$ M	approximately 4×10^{-4} M

aldehyde system (CH=CHCHO) can evidently be raised considerably by at least a factor of ten by the introduction of the polar hydroxyl group at C-4. Substance (VI) therefore occupies second place in importance, in respect of biological and biochemical interest, to those substances isolated from preparation LHPO. Since 4-hydroxyoctenal (HOE) was easily isolated in larger amounts, and since it has been obtained by Esterbauer *et al.* on a large scale by several synthetic routes (cf section 3.2.3), the investigations on the biochemical and biological properties of this substance presented no further difficulties (cf the number of references cited in line 6 of table 3.7).

3.2.2 Biochemical and biological effects of natural 4-hydroxy-trans-2,3-octen-1-al (HOE)

Table 3.7 gives information on the inhibition of anaerobic glycolysis and respiration of native EATC by HOE. Figure 3.4 presents graphically the other effects on EATC with HOE. One can see that HOE inhibits respiration only in concentrations which are able to block anaerobic glycolysis completely. This behaviour appears characteristic for genuine glycolysis poisons, such as hydrogen peroxide, X-rays, monoiodoacetate, ethyleneiminoquinones. However, the action of HOE differs from these glycolysis poisons in one significant respect. The inhibition of glycolysis

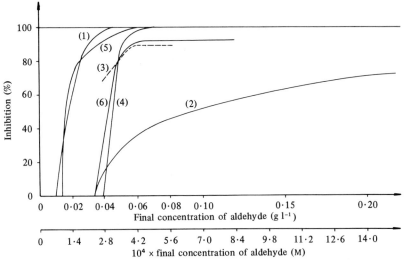

Figure 3.4. Dose-response curves for the inhibitory effects of HOE. Curves (1), (3), and (4) show inhibition of anaerobic or aerobic glycolysis, and respiration in Ehrlich ascites tumour cells, respectively, after aerobic preincubation of the cells. Curve (2) shows anaerobic glycolysis after anaerobic preincubation of the cells. Curves (5) and (6) show inhibition of GAPDH and LDH, respectively, extracted from the cells after they had been incubated with HOE (according to Schauenstein[130]).

occasioned by these typical poisons can be reversed by addition of NAD during the preincubation, but this is not the case with HOE[150]. This suggests that HOE does not inhibit anaerobic glycolysis by blocking the NAD system, and this can be seen clearly in figure 3.5. Although, like 2,3,5-triethyleneiminobenzoquinone or hydrogen peroxide, HOE can, with increasing concentration, lower the cellular NAD to 15% of normal, the curve of this lowering of NAD shows a quite different situation[151]. The points marked x in figure 3.5 show that addition of nicotinamide in part makes up for the decrease in NAD, and yet the inhibitions of anaerobic glycolysis and of respiration remain fully intact. One may largely explain these findings by considering curves (5) and (6) in figure 3.4. It clearly follows from these curves that the same HOE concentration, which completely blocks anaerobic glycolysis, can also completely deactivate one of the most important glycolytic enzymes, glyceraldehyde phosphate dehydrogenase (GAPDH), and deactivate the lactate dehydrogenase (LDH) by 50%. The inhibition of glycolysis in EATC may not be due exclusively to the deactivation of the GAPDH, but this process will certainly contribute substantially to it. Conclusions have not yet been possible on the mechanism of action of HOE as an inhibitor of respiration. Nevertheless the deactivation of the two enzymes provides important insight into the basic mechanism of the activity of hydroxyenals. Initially it was shown that the inhibition of crystallized enzymes by HOE could be removed by addition of a one hundredfold excess of cysteine (Cys)[130, 152].

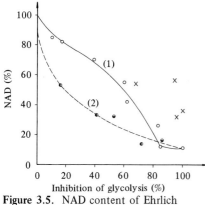
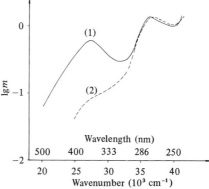

Figure 3.5. NAD content of Ehrlich ascites tumour cells compared with the degree of anaerobic glycolysis. All values were obtained after aerobic incubation (according to Schauenstein[130]). ○, HOE in buffer without glucose; ◐, H_2O_2; ●, Trenimon without glucose; x, HOE without glucose but with added nicotinamide.

Figure 3.6. Ultraviolet absorption spectrum of glyceraldehyde dehydrogenase: (1) after preincubation in HOE, and subsequent treatment with 2,4-dinitrophenylhydrazine, and exhaustive dialysis; (2) after treatment with 2,4-dinitrophenylhydrazine (according to Dorner[153]). $m = E/d$; $E = \lg(I_0/I)$; d: in cm.

Of the original solution of crystallized GAPDH from rabbit muscle (Boehringer Mannheim GmbH), 0·5 ml was incubated with 3·97 mg of HOE in a total volume of 2 ml, at 2°C for thirty minutes. This was followed by dialysis at 2°C until equilibrium was attained. The dialysate was then treated with 0·06 ml of a solution of 2,4-dinitrophenylhydrazine (DNPH), and the mixture allowed to react for sixteen to eighteen hours. This was again followed by dialysis to remove completely the excess of DNPH, and the dialysed enzyme solution was then measured spectrometrically, giving curve (1) of figure 3.6[24]. The absorption maximum at 36 000 cm^{-1} (278 nm) corresponds to the enzyme protein; that at 27 300 cm^{-1} (366 nm) shows convincingly the presence of a protein-bound 2,4-DNPH hydrazone. GAPDH, without prior treatment with HOE, when treated with DNPH shows only a very weak shoulder in the region of 27 300 cm^{-1} (curve 2, figure 3.6). The following conclusions may then be drawn from curve 1 of figure 3.6:

(1) HOE has entered into chemical binding with the enzyme protein;
(2) with this chemical binding to the protein the aldehyde group has clearly remained in a reactive form, giving with DNPH the corresponding hydrazone derivative;
(3) the wavelength of the hydrazone maximum shows that one is dealing with a hydrazone of a saturated aldehyde grouping;
(4) the quantitative assessment of the intensity of the hydrazone maximum gives on average 4·09 ± 0·04 hydrazone residues, indicating that four HOE molecules are bound per enzyme molecule;
(5) this corresponds to the number of sulphydryl groups of the GAPDH responsible for enzyme activity;
(6) in the original incubation, where the molar ratio used for enzyme/HOE was 1:800, the enzyme activity completely disappeared. It could be fully restored with an excess of Cys (see above); it may thus be concluded that HOE reacts selectively, under the given conditions, with only the four functional sulphydryl groups of the total of sixteen sulphydryl groups of the enzyme, probably by linking to the 2,3-carbon–carbon double bond of HOE.

This interpretation was completely confirmed by chemical and physical analyses of the product from a model reaction carried out between glutathione (GSH) and HOE. The constitution of the product could unambiguously be identified as follows[25]:

$$\begin{array}{c} \text{Glu Cys Gly} \\ | \\ \text{S} \\ | \\ \text{CH}\!-\!\text{CH}_2 \\ \text{CH}_3(\text{CH}_2)_3\text{CH} \quad \text{CHOH} \\ \diagdown\text{O}\diagup \end{array}$$

The reaction between a sulphydryl group and HOE thus consists of an initial addition to the 2,3-carbon–carbon double bond, with formation of a saturated aldehyde, which immediately cyclizes to an internal hemiacetal. Treatment with the acidic DNPH reagent opens up the ring and allows the aldehyde group to react with the hydrazine reagent.

In this way the basic reaction mechanism of all 4-hydroxyenals (HE) in biological and biochemical systems has been experimentally demonstrated in an unambiguous manner, and subsequent investigations have without exception confirmed this interpretation.

In significant contrast to the other aldehydes considered here, the hydroxyenals do not react with any other group, such as an amino or similar group, even at an approximately comparable rate. One can therefore be certain that all the biochemical and biological effects occasioned by HE, known up to the present time, arise through reactions with functional thiol groups. Details of these effects will be referred to when 4-hydroxypent-2-en-1-al is discussed.

The rate with which a sulphydryl group reacts with HE is determined by its reactivity, which in turn depends on the size, shape, and chemical constitution of the molecule which bears the sulphydryl group. In general, thiol compounds of lower molecular weight react appreciably faster with HE than do those of higher molecular weight; thus, Cys and GSH react faster than sulphydryl proteins. With proteins the primary, secondary, or tertiary structure is decisive for the reactivity of the thiol groups; these factors also determine the higher reactivities of sulphydryl groups in enzymes. Functional thiol groups show substantial differences in their chemical reactivities. Thus one can understand why HOE reacts with different sulphydryl enzymes at very different rates, and thereby deactivates these enzymes to very different degrees, as is shown in table 3.9. A detailed analysis of the kinetics of thiol groups with HE is given in section 3.1.2.

Table 3.9. Inhibitions, caused by HOE, of the activities of several crystallized sulphydryl enzymes of glucose metabolism (according to Schauenstein[130]).

Enzyme	Final concentration of HOE (mM)	Inhibition (%)
GAPDH	0·03	50
	0·10	100
LDH	8·0	50
	8·5	64
ALD	8·5	0
HK	8·5	0
G 6-PDH	8·5	0

The fact that the same enzymes or enzyme systems present in different types of cells are inhibited by HE in very different ways is perhaps more interesting from the biological point of view. This finding, which is characteristic for all HE, was already clearly apparent for HOE, the first representative discussed here. The figures given in table 3.10 show[155] that HOE does not measurably influence the respiration of normal liver and kidney cells, even under extreme altered conditions.

With an HOE concentration in the region $\geqslant 3$ mM characteristic morphological cell damage occurs after periods of $\geqslant 10$ minutes. Protrusion of circumscribed cytoplasmic droplets takes place on the cell surface, and these droplets subsequently become separate and independent.

Schauenstein et al.[148] first observed this effect with HOE, but found it also with other cytotoxic agents, such as hydrogen peroxide, dilute

Table 3.10. Differences in inhibitions, caused by HOE, of the oxygen consumption of EATC, and of monkey liver and kidney cells (according to Schauenstein et al.[148] and Jaag[151]). Results were obtained after 30 minutes preincubation in a total volume of 2·5 ml.

Number and type of cells	Inhibition of oxygen uptake (%)	Final concentration of HOE (M)
$1·5 \times 10^7$ EATC	90–95	$\geqslant 3·9 \times 10^{-4}$
$2·5 \times 10^5$ monkey liver cells	0	$\geqslant 4·8 \times 10^{-4}$
3×10^5 monkey kidney cells	0	$\geqslant 1 \times 10^{-2}$

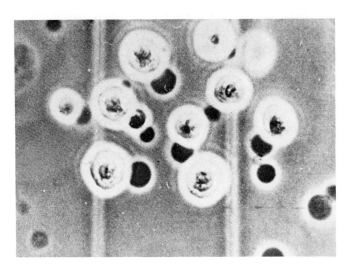

Figure 3.7. Ehrlich ascites tumour cells (6th–8th day *post implantationem*) after 30 minute aerobic incubation in 5 mM HPE, at 37°C (phase-contrast photograph of Schauenstein et al.[148]).

hydrochloric acid, and various cytostatic agents. It was also described after exposure to X-rays[156]. Ratzenhofer and Zangger[157] termed this protrusion 'stalagmosis', and the detachment of the droplets 'stalagmoptysis', and found that the latter indicated the death of the Ehrlich ascites tumour cell, as shown by the Trypan Blue coloration and the inhibition of cell multiplication in the living animal[148] (figure 3.7).

Hence we meet for the first time a genuine biological effect of HOE, which appears to be related to the specific biochemical effects previously discussed.

It is worth noting that no morphological damage could be detected in cultured rat liver and kidney cells with concentrations of HOE up to 20 mM [19]. This is in agreement with the absence of inhibition of energy metabolism (see table 3.10).

3.2.3 Syntheses of 4-hydroxyenals

The biochemical and biological properties of 'natural' HOE, referred to above, make it desirable to have available larger quantities of this substance, for use in extensive investigations. Since the extraction from preparation LHPO is tedious and the yields are low, possibilities of a synthetic preparation were explored. Esterbauer and Weger[158] initially used a procedure that was somewhat cumbersome and was not satisfactory in respect of yield. Later they developed a simpler synthesis[159], which gave significantly better yields (scheme 3.7). It is worth noting that hydrogenation of the carbon–carbon triple bond over Lindlar catalyst, which is commonly used for this purpose, leads to a carbon–carbon double bond in a *cis* configuration; this immediately leads to cyclic acetal

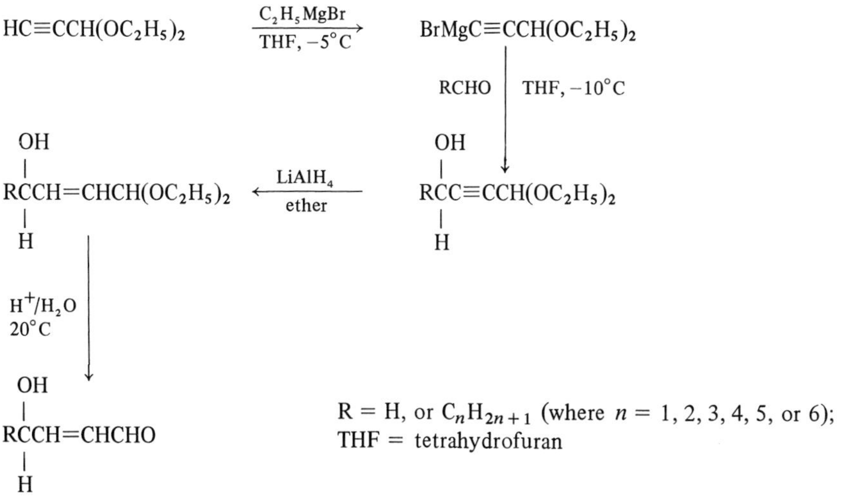

$$HC\equiv CCH(OC_2H_5)_2 \xrightarrow[THF, -5°C]{C_2H_5MgBr} BrMgC\equiv CCH(OC_2H_5)_2$$

$$\xrightarrow[THF, -10°C]{RCHO}$$

$$RCCH=CHCH(OC_2H_5)_2 \text{ (OH, H)} \xleftarrow[ether]{LiAlH_4} RCC\equiv CCH(OC_2H_5)_2 \text{ (OH, H)}$$

$$\xrightarrow{H^+/H_2O, 20°C}$$

$$RCCH=CHCHO \text{ (OH, H)}$$

R = H, or C_nH_{2n+1} (where n = 1, 2, 3, 4, 5, or 6); THF = tetrahydrofuran

Scheme 3.7

formation, namely to dihydrofuran derivatives, which cannot be converted into the desired 4-hydroxyalkenals (scheme 3.8). With the above synthesis (scheme 3.7) the means were found for the synthesis of homologous HE and a multitude of derivatives[160]. The foundations were thus laid for a systematic investigation to be made.

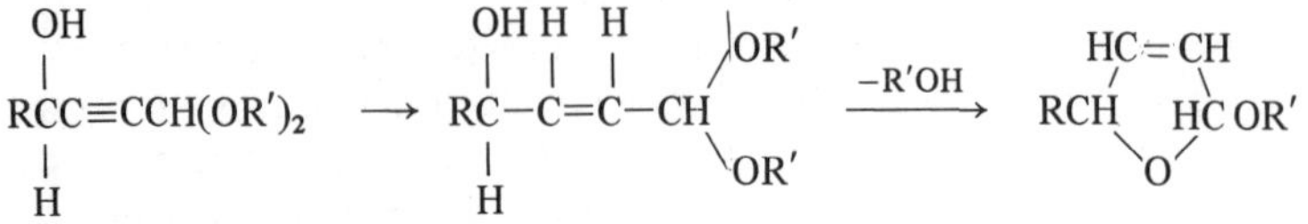

$$\underset{\underset{H}{|}}{\overset{\overset{OH}{|}}{RCC\equiv CCH(OR')_2}} \longrightarrow \underset{\underset{H}{|}}{\overset{\overset{OH\ H\ H}{|\ \ |\ \ |}}{RC-C=C-CH}}\!\!\!\diagdown_{OR'}^{OR'} \xrightarrow{-R'OH} \underset{\diagdown_{O}\diagup}{\overset{HC=CH}{RCH\ \ \ HCOR'}}$$

Scheme 3.8

3.2.4 Biochemical and biological effects of synthetic 4-hydroxyenals

First of all it was established that natural and synthetic preparations of HOE possessed the same biochemical and biological activities, after which synthetic HOE was used exclusively. Two series of investigations, in particular, deserve mention:

First Ardenne[161] found that HOE under hyperthermia (42°C) highly and selectively sensitizes EATC compared with liver cells. In comparative investigations with disulphiram and vitamin K_3 results were obtained which are reproduced in table 3.11.

Second Burk and Woods[162] examined the action of HOE on mouse melanoma S91 and EATC in combination with hyperthermia *in vitro* and *in vivo*. They described HOE as one of the most active and most selective anticancer agents that they had studied in the last twenty-five years. In connection with the selective action on energy metabolism and the viability of tumour cells *in vitro* they confirmed not only our own findings[148], but in addition found selective damage after intravenous or intraperitoneal injection at locations far removed from the site of the intramuscularly implanted melanoma. Damage was also found in intraperitoneally implanted EATC after intraperitoneal application of HOE. Thus melanoma S91, which was implanted in the leg muscle and

Table 3.11. Selective sensitization of EATC with disulphiram, HOE, and vitamin K_3 (according to Ardenne[161]).

Property	Units	Disulphiram	HOE	Vitamin K_3
Concentration needed for good sensitization	g/ml	2×10^{-5}	$6 \cdot 5 \times 10^{-5}$	$\leqslant 3 \times 10^{-5}$
Maximal tolerated intraperitoneal dose	g/kg mouse	1	$6 \cdot 5 \times 10^{-2}$	$\sim 8 \times 10^{-2}$
Solubility in water at 37°C	g/ml	5×10^{-6}	$1 \cdot 3 \times 10^{-3}$	3×10^{-1}
Selectivity		good	very good	very good

was excised one hour after being treated, intravenously or intraperitoneally, with a single dose of 1-2 mg of HOE, showed in the Warburg respirometer a 30%-50% inhibition of anaerobic glycolysis, respiration, and aerobic glycolysis. Values from melanoma of untreated animals were used for comparison. The similarly excised kidney and liver of the HOE-treated animals did not show any metabolic defects. Similar effects were found with intraperitoneally inoculated EATC after intraperitoneal injection of 2 mg of HOE, followed by considerable increase of the percentages of Trypan Blue-positive and morphologically irreversibly damaged (that is, dead) cells. Only with much higher HOE dosages did damage occur also in liver and kidney. HOE acts, according to Burk and Woods, not only synergistically with hyperthermia *in vitro*, as already reported by Ardenne[161], but also *in vivo*. Selective damage of cancer cells is greater *in vivo* than *in vitro*. It is also greater *in vivo* with EATC older than 4-5 days *post implantationem*, presumably because the younger cells have a larger supply of glycose and/or oxygen.

The synthesis of 4-hydroxyenals opened up for the first time the possibility of comparing the biochemical and biological activities of homologous HE of different chain-lengths. This is a problem which we also encounter for example in the series of investigations of Szent-Györgyi *et al.* with methylglyoxal derivatives (see section 5.1.5.2).

Available substances were the five-, six-, seven-, and eight-carbon atom homologous HE, as synthesized by Esterbauer and Weger[158-160]. The test system was EATC in the mouse. An isotonic solution (1 ml) of different HE was injected subcutaneously into the left hind flank of the experimental animal, whereupon a swelling formed at the injection site. Into these swellings 10^7 EATC (6th-8th day *post implantationem*) in $0\cdot1-0\cdot2$ ml of ascites serum were injected. At 30 minutes after this 'incubation *in vivo*' the tumour cells were isolated, and the percentage of killed cells was determined with Trypan Blue. It was shown that all the HE compounds investigated killed off with certainty a maximum of 90%-95% of the implanted tumour cells *in vivo*. In this the five-, seven-, and eight-carbon atom compounds were practically of equal effectiveness, whereas the six-carbon atom compound was appreciably less active. Thus approximately 80% of cells were killed by $3\cdot5 \times 10^{-7}$ mol/g body weight of five-, seven-, and eight-carbon atom compounds, whereas the six-carbon atom compound was successful only to approximately 50%. With smaller doses the difference in effective strength is even more marked[163].

Kollaritsch[164] recently showed in detailed investigations carried out at the Institut für Biochemie der Universität Graz how strong an influence the length of the aliphatic chain residue has on the chemical and biological activity of hydroxyenals. For these experiments the higher homologous hydroxyenals, as synthesized by Esterbauer[158-160], were made available.

Table 3.12 gives the results obtained. It was shown that the inhibitory action of hydroxyenals on the incorporation *in vitro* of [^{3}H] thymidine into the DNA of EATC parallels the dependence on the chain-length of the cytotoxic action on implanted tumour cells *in vivo*. In particular one should note the activity minimum with the six-carbon atom homologue, and the marked increase in activity from the nine-carbon atom compound upwards.

The toxicity of the homologous hydroxyenals shows a similar behaviour, only in this case minimal activity is reached less sharply and is shifted to the eight-carbon atom aldehyde. The five-carbon atom homologue possesses the important advantage for biological experiments of improved water solubility, when compared with the higher homologues. Since the four-carbon atom aldehyde could not be considered, owing to the exceptional difficulty of its synthesis in substantial amounts, further investigations were carried out with the five-carbon atom homologue, 4-hydroxypentenal (HPE).

Table 3.12. Chemical and biochemical properties of homologous hydroxyenals (according to Kollaritsch[164]).

Chain-length	C_{50} (mol × 10^{-8})	LD_{50} (mol × 10^{-4})	W	K
C_4			360	12·8
C_5	4·8	9·9	132	4·5
C_6	8·1	9·8	60	1·0
C_7	5·6	12·2	30	0·3
C_8	4·6	14·6	12	0·1
C_9	2·5	4·4	7	0·04
C_{10}			3	0·01
C_{11}	1·2	4·6		0·004
C_{14}	1·8			
C_{15}	1·6			
C_{17}	1·7			

C_{50}: concentration required for half-maximal inhibition of [^{3}H]thymidine incorporation into native EATC (6th–8th day *post implantationem*), in mol/10^6 cells.
LD_{50}: determined for NMRI mice, 20–25 g, of both sexes, in mol/kg body weight.
W: solubility in water at 20°C, in g/litre. K: partition coefficient, water–chloroform. For the determination of the C_{50} and LD_{50} values for the fourteen, fifteen, and seventeen carbon atom hydroxyenals, ethanol had to be added to the incubation mixtures in amounts of 10–50 μl/2 ml.

3.2.5 *Biochemical and biological action of 4-hydroxypentenal (HPE)*
3.2.5.1 Involvement in energy metabolism and macromolecular biosyntheses
Table 3.13 gives information on the results obtained with EATC for increasing HPE concentrations *in vitro*. It can be seen that HPE actively inhibits glycolysis and respiration, as could be expected from the earlier experience with HOE. Aerobic glycolysis, results for which are not presented here, proved less susceptible to HPE than anaerobic glycolysis.

Table 3.13. Action of different concentrations of HPE on the metabolism of EATC (according to Bickis et al.[165]). Concentration of glucose: 10 mM. After gassing and thermal equilibration, tracer and HPE were tipped from the side arm of the Warburg flask into the main well, and measurements were begun. Metabolic activities are given per 50 mg of cells/h with values in parentheses for the last 20 minutes.

1 h aerobic incubation (air)

concentration of HPE (mM)	oxygen uptake (μl)	Leu into protein (nmol)	A into RNA (nmol)	A into DNA (nmol)
0	36 (12)	33·3	9·4	1·73
0·1	35 (12)	23·4	8·6	0·69
0·2	28 (9)	11·1	5·6	0·14
0·3	24 (8)	9·4	4·7	0·08
0·4	20 (5)	7·1	3·7	0·08
0·5	17 (3)	6·7	2·8	0·05

1 h anaerobic incubation ($N_2 + CO_2$, 95:5)

concentration of HPE (mM)	glycolysis (μl of CO_2)	Leu into protein (nmol)	A into RNA (nmol)	A into DNA (nmol)
0	278 (79)	40·1	13·2	2·15
0·1	291 (86)	22·5	10·3	0·57
0·2	231 (62)	12·7	7·0	0·15
0·3	208 (52)	9·6	5·0	0·07
0·4	191 (47)	8·1	4·1	0·05
0·5	180 (41)	6·4	3·3	0·04

Table 3.14. Time dependence of the inhibitory activities of HPE on EATC on anaerobic incubation (according to Bickis et al.[165]). Values in parentheses are for the last 20 minutes.

Duration of incubation (min)	Anaerobic glycolysis (CO_2)		Leu into protein	
	blank	HPE (0·2 mM)	blank (nmol)	HPE (0·2 mM)
30	200 (97)	193 (87)	19·1	7·1
60	381 (87)	331 (65)	33·8	10·0
90	541 (76)	408 (47)	48·6	11·6
120	672 (73)	500 (38)	60·0	13·5

	A into RNA		A into DNA	
	blank (nmol)	HPE (0·2 mM)	blank (nmol)	(HPE 0·2 mM)
30	5·9	3·4	0·79	0·16
60	9·1	4·8	1·54	0·17
90	11·9	5·4	2·33	0·18
120	14·3	6·4	3·11	0·18

Table 3.15. Comparison of the effects of HPE and other carcinostatic agents on human tumours *in vitro* (according to Bickis et al.[165]). Aerobic incubation was for 2 h in Krebs–Ringer bicarbonate glucose. The labelled substances were $[1\text{-}^{14}C]$leucine and $[8\text{-}^{14}C]$adenine. All the inhibitory substances were added to the tissue slices 20 minutes before gassing and thermal equilibration. The figures in the table represent activities as percentage of the control values.

Tumour	Biosynthesis		HPE (0·3 mM)	Trenimon (4 μM)	Fluorouracil (3·5 mM)	Methotrexate (0·1 mM)	VLB (0·06 mM)	Mitomycin C (0·15 mM)
Lymph node metastasis of an oviduct carcinoma	protein:		73	100	92	100	84	88
	RNA:		77	98	71	116	98	67
	DNA:		45	43	42	27	72	79
Hidr-adenocarcinoma	protein:		43	66	92	94	58	58
	RNA:		54	64	73	105	67	56
	DNA:		46	55	54	45	37	40
Metastasis in the omentum of a colon carcinoma	protein:		29	—	63	70	—	—
	RNA:		35	—	61	77	—	—
	DNA:		42	—	77	73	—	—

It was surprising to find that the incorporation of radioactive precursors into DNA, proteins, and RNA reacts much more sensitively to HPE than does the energy metabolism. With as little as 0·1 mM HPE, at which concentration respiration and glycolysis are not yet measurably affected, the incorporation of adenine into DNA is already inhibited by approximately 60%, and the incorporation of leucine into proteins by approximately 30%. Both inhibitions increase to 90% and 70% respectively with 0·2 mM HPE.

The synthesis of RNA, on the other hand, appears to be less sensitive, though 70%–75% inhibition is achieved with 0·5 mM HPE. Virtually the same results were achieved with ascites sarcoma-37 cells. The greater susceptibility of the synthesis of DNA is also clearly shown in the early development of the different inhibitory actions at constant HPE concentration (table 3.14).

The data show that with 0·2 mM HPE the DNA synthesis has probably ceased within the first 10 minutes, whereas anaerobic glycolysis has not yet been measurably affected. The inhibitions of RNA and protein syntheses are also not complete after incubation for two hours. Because of the drastic inhibition of biosyntheses caused by HPE in EATC and sarcoma-37 cells in the mouse, it seemed of interest to compare the action of HPE *in vitro* on the respiration and biosyntheses of primary tumours in man, with those of other cytostatic agents. The figures of table 3.15 show that the action of HPE on the metabolism of three human tumours *in vitro* is quite comparable with the effects of five cytostatic agents already in clinical use. Especially surprising is the high HPE sensitivity of the DNA synthesis of the colon carcinoma, which is manifestly resistant towards methotrexate and 5-fluorouracil. Both these cytostatic agents are specific inhibitors of DNA synthesis, and usually are among the most effective agents for gastrointestinal tumours[166].

3.2.5.2 Inhibition of glycolysis; action mechanisms of the observed inhibitions

As expected from the results with HOE the inhibition of glycolysis caused by HPE is also not abolished by addition of nicotinamide[165]. The inhibition may therefore be similarly caused essentially by the deactivation of the GAPDH. This enzyme is also deactivated by HPE with high selectivity in model experiments in comparison with five other sulphydryl enzymes (hexokinase, G-6-PDH, aldolase, LDH, and alcohol dehydrogenase)[165]. The selective inhibitory activity of HPE on GAPDH is also illustrated by the fact that HPE does not deactivate the dehydrogenases of the pentose phosphate cycle in EATC[165].

The deactivation of crystallized GAPDH by HPE is connected with the loss of about three sulphydryl groups, or the binding of about four molecules of HPE per molecule of enzyme[165,167]. The deactivation of crystallized GAPDH may be completely reversed within seconds by Cys[165], and so it follows that HPE deactivates the GAPDH through reaction with the functional sulphydryl groups. The same mechanism may be assumed to occur also in the cell, since the inhibition of glycolysis caused by HPE in EATC and sarcoma-37 cells can be completely prevented and abolished by addition of GSH or Cys[165]. Since GAPDH activity appears to be the step in the glycolysis chain most sensitive to HPE, it can be readily appreciated how HPE can disturb the NAD^+/NADH levels in the cytosol. This is influenced to different degrees by different sulphydryl reagents and other glycolysis inhibitors. The inhibitions may be prevented in different ways, or abolished, as table 3.16 shows.

Table 3.16. Comparison of the mode of action of various sulphydryl reagents and cytostatic agents (according to Bickis et al.[165]).

Inhibitor	Inhibition of GAPDH-catalysed reactions may be	
	abolished by	prevented by
p-Chloromercuribenzoate[168]	Cys	high substrate concentration
Iodoacetate[168,169], iodobenzoate	not effective	not effective
Triethyleneimine[168,170,171] carcinophilin	NA	NA
HPE	GSH, Cys	GSH, Cys

3.2.5.3 Inhibition of respiration

The factors responsible for the inhibition by HPE of glucose respiration, are as follows.
(1) The deactivation of GAPDH.
(2) The drastic lowering of the cytoplasmic NAD level, by approximately 70%[172].
(3) The fact that addition of pyruvate, glutamate, succinate, and fumarate, singly or in combination, is unable to remove the inhibition[165], pointed clearly, and for the first time, to a blocking of oxidation in the tricarboxylic acid cycle and/or in the respiration chain.

Schauenstein and Kapfer[173] isolated from HPE-treated EATC the mitochondrial malate dehydrogenase and the isocitrate dehydrogenase, which were deactivated to approximately 30% and 40% respectively, compared with the enzymes of the control cells. It was also found that with HPE-treated EATC an accumulation of succinate occurs of approximately 95%, compared with the control cells, and there is a corresponding drastic reduction of the cytochemical staining with 3-(4,5-dimethylthiazol-2-yl)-2,5-diphenylmonotetrazolium bromide (MTT reagent) as an indicator of the succinate dehydrogenase (SDH) activity. We cannot deduce with certainty that a deactivation of SDH must occur, but we can conclude that succinate oxidation is strongly inhibited, and further evidence is provided how inhibition of respiration through HPE is realized.
(4) If NADH is added as substrate to EATC, which had been treated with 1 mM HPE, thereby reducing respiration of the cells to 75%, then the inhibition of respiration returns to a moderate 47%.

With 3 mM HPE solution the inhibition of respiration, under the same experimental conditions, is decreased from over 90% to 60% by the addition of NADH[172]. These results indicate that the respiratory chain is also blocked by HPE and that this effect contributes substantially to the inhibition of oxygen uptake.

As was shown in the discussion of the inhibition of glycolysis (table 3.16), the mechanism of action by HPE in energy metabolism is significantly different from those of other sulphydryl inhibitors. This difference is further demonstrated by the following results. HPE inhibits glycolysis and respiration at practically the same concentration. Iodoacetate, DL-glyceraldehyde, and carcinophilin, on the other hand, stimulate respiration (at least with tumour cells) at concentrations that cause nearly complete inhibition of the glycolysis, and inhibit respiration only at much higher concentrations[174]. Further, the oxidation steps of the pentose phosphate cycle are inhibited, in marked contrast to HPE, by *p*-chloromercuribenzoate[175], and by iodoacetate[176].

3.2.5.4 Inhibition of nucleic acid biosyntheses

Valuable contributions to this topic have been made by Seeber et al.[177]. Thus, they found with EATC that HPE at concentrations as low as 50 μM decreases the incorporation of [^{14}C]thymidine and [^{14}C]uridine into DNA—the incorporation of [^{14}C]thymidine is particularly strongly inhibited. The total activity was then measured of the nucleosides and nucleotides in the acid-soluble fraction. At 1 mM HPE a decrease in the total activity to 55% of the control value is observed (on average), clearly indicating an inhibition of nucleoside entry into tumour cells.

Subsequently the different nucleoside phosphates of the acid-soluble fraction were separated, and quantitatively determined. A definite accumulation of mono- and diphosphates was observed, with an accompanying decrease of the triphosphates (table 3.17).

Table 3.17. Influence of HPE (1 mM) on the percentage distribution of mono-, di-, and triphosphates of different nucleosides (according to Seeber et al.[177]).

^{14}C-labelled nucleoside		Triphosphate	Mono- and diphosphate
Thymidine	control	87	13
	HPE	42	58
Uridine	control	63	37
	HPE	34	66
Adenosine	control	80	20
	HPE	46	54
Deoxyadenosine	control	72	28
	HPE	38	62

The DNA-polymerase and the DNA-dependent RNA-polymerase from EATC were examined in a cell-free system for sensitivity towards HPE. Both enzyme systems are inhibited by HPE. On addition of HPE to the whole mixture (final concentration of HPE 0·7 mM and 0·13 mM) immediately before starting the experiment only a 20% inhibition takes place with the DNA-polymerase, but with RNA-polymerase practically no inhibition at all occurs. On preincubation (40 minutes, 22°C) of the enzyme fraction with HPE, the DNA-polymerase shows, with the above-mentioned HPE concentrations, inhibitions of 62% and 77% respectively, and with the same preincubation RNA-polymerase with 0·7 mM HPE shows only a 46% inhibition. On preincubation of the primer DNA with HPE, at both concentrations a 30% inhibition occurred subsequently with the DNA-polymerase and 10% inhibition of RNA-polymerase with 0·7 mM HPE.

In summary, then, the measurements of Seeber *et al.* show that the complete inhibitions of DNA and RNA biosyntheses, as established for intact EATC, may be explained by the following effects:
(1) Inhibition of nucleoside phosphorylation results in an insufficient supply of substrates for the polymerases.
(2) A direct attack of HPE on the polymerases themselves.
A major influence on the matrix function of primer DNA is, however, unlikely.

3.2.5.5 Reaction with sulphydryl groups
The results referred to so far have shown that HPE inhibits the biosyntheses of nucleic acids and proteins, and energy metabolism in EATC and cells of sarcoma-37 of the mouse. All inhibitory effects can be prevented as well as abolished by addition of cysteine.

The metabolic paths concerned are controlled by sulphydryl enzymes. Since HPE is a highly active sulphydryl reagent, it appears plausible to explain the effects reported in terms of a reaction of HPE with the functional sulphydryl groups of the corresponding enzymes in the manner described (cf pages 47–49). The enzymes are thus deactivated and the corresponding metabolic pathways are inhibited.

This reaction takes place via an equilibrium; since HPE reacts more rapidly with lower rather than with higher molecular weight sulphydryl compounds (cf page 49), the addition of appropriate amounts of Cys or GSH causes them to react directly with the free HPE of the equilibrium. The equilibrium is thereby constantly disturbed, free HPE being furnished by the compounds (II) and (III) of scheme 3.9. A reactivation of the enzyme results.

This interpretation is treated in more detail in the section on the kinetics of the reaction between α,β-unsaturated aldehydes and sulphydryl groups (section 3.1.2). If this view is correct the following must be expected for the reversible interaction between HPE and protein sulphydryl groups:

(a) *A decrease in free sulphydryl groups.* This effect may be expressed for pure proteins of known molecular weights as $-\Delta$PSH/mol, or alternatively as $-\Delta$PSH/g of protein, where PSH is the protein thiol. In the reversible reaction between HPE and whole cells, the former can react with the sulphydryl groups of soluble cell proteins (PSH_s), or of insoluble structural proteins (PSH_i), and nonprotein thiol compounds (NPSH). In the first case the loss of sulphydryl groups may be expressed as $-\Delta$PSH/g of soluble cell proteins or per g wet weight, or per a specific number of cells. Microspectrophotometrically the sulphydryl groups of trichloroacetic acid-insoluble proteins are determined as $PSH_{tot.}$ ($= PSH_s + PSH_i$). The loss of $PSH_{tot.}$ ($-\Delta PSH_{tot.}$) in determinations on a preparative scale may be expressed either per g wet weight or per single cell. In microspectrophotometric determinations values are usually expressed per single cell.

Loss of NPSH is given as $-\Delta$NPSH per g wet weight, or for a specific number of cells.

(b) *A corresponding binding of HPE to the protein.* With isolated pure proteins the data may again be expressed per mol, or per g of bound HPE molecules. With whole cells the data may be given as the number of bound HPE molecules per g wet weight, or, when determined cytospectrometrically, per single cell. The numbers given in table 3.18 show that both criteria of the purest proteins are fulfilled[167]. In particular the results obtained with GAPDH and ovalbumin show us that HPE undergoes reaction only with highly reactive sulphydryl groups. GAPDH possesses a total of about ten sulphydryl groups, of which four are functional groups.

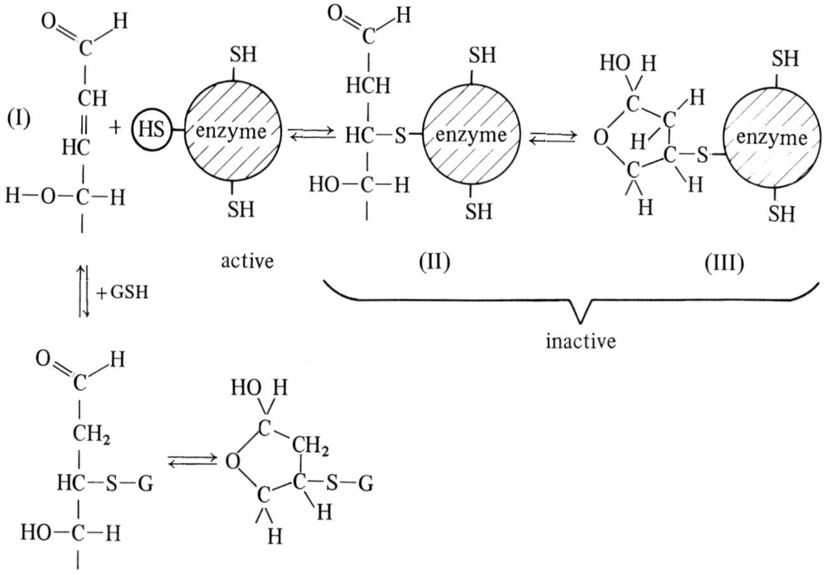

Scheme 3.9

It is known that on binding with about four molecules of HPE the enzyme is completely deactivated (cf page 48). Ovalbumin possesses altogether four sulphydryl groups, which, according to the literature, are sluggish in their reactivity, or in part masked by being built into the inner region of the molecule. HPE reacts with none of these groups.

According to data in the literature, aldolase possesses altogether twenty-eight to twenty-nine sulphydryl groups, only half of which are detectable by means of 5,5'-dithio-bis-(2-nitrobenzoic acid) (DTNB). Four sulphydryl groups are responsible for the enzymatic activity, but are nevertheless of low reactivity[178]. After treatment with HPE, aldolase binds 1 mol of HPE/mol, and thereby loses one sulphydryl group. This group should belong to the more reactive sulphydryl groups, which is not, however, responsible for the enzyme activity. Therefore, the enzyme (as shown earlier, pages 49, 57) is not deactivated by HPE at these concentrations.

The figure of 0·7 mol of HPE bound/mol of BSA corresponds exactly to the number of sulphydryl groups determined amperometrically with silver nitrate. The lower value found with DTNB is explained by the fact that the determinations were always carried out without using a detergent, whereas in some cases sulphydryl groups become available to DTNB only with detergent.

Reduction of S—S bonds in cysteine evidently furnishes highly reactive sulphydryl groups, presumably by opening up the tertiary and quaternary structures of the protein molecules. All the free sulphydryl groups thus obtained react promptly with HPE. The number of HPE molecules bound to the protein does not correspond in every case exactly to the number of sulphydryl groups which have disappeared at the same time. This is readily explained by the basically different methods of determination used for both quantities, for which, nevertheless, an unambiguous linear relationship exists. A sample correlation coefficient obtained from all six value pairs was calculated to be 0·9977. This proves conclusively that HPE is bound to the sulphydryl groups of the proteins.

Table 3.18. Comparison of the number of molecules of HPE bound per molecule of protein, of the sulphydryl groups consumed as a result, and of the thiol groups determined with 5,5'-dithio-bis-(2-nitrobenzoic acid) (DTNB). The binding with HPE was determined by means of 2,4-dinitrophenylhydrazine (see page 48).

Protein	HPE bound (mol/mol)	$-\Delta PSH_s$ (mol/mol)	PSH (total/molecule)
GAPDH	4·63 ± 0·07	3·20 ± 0·11	9·22 ± 0·38
Aldolase	0·84	1·1	14·6
BSA	0·70	0·42	0·47 ± 0·03
BSA (red.)[a]	24·4	24·9	25·3 ± 1·6
β-Lactoglubulin (red.)[a]	9·3	8·72 ± 0·27	9·1 ± 0·34
Ovalbumin	0	0	4

[a] red. = after reduction with thioglycollate (according to Schauenstein et al.[167]).

The reactions between HPE and crystallized pure proteins described here can serve as models for the reactions of HPE in intact living cells. This will be shown initially with the example of EATC.

First for NPSH and PSH_s. Cells (six days *post implantationem*) were preincubated in isotonic 5 mM HPE solution for thirty minutes at 37°C (controls in 0·9% sodium chloride solution), and then homogenized under nitrogen. After removal of the insoluble components, the homogenate was separated on Sephadex into a high and a low molecular weight fraction (figure 3.8)[179]. Fraction A contains the PSH_s, and fraction B the NPSH. The sulphydryl groups were determined for these solutions with DTNB, and for fraction A the protein concentrations were also determined. The values obtained are given in table 3.19.

Second for PSH_{tot}. EATC were treated with HPE (5 mM) (controls with 0·9% sodium chloride solution), appropriately fixed, and stained for PSH_{tot}. with dihydroxydinaphthyl disulphide (DDD reagent) according to the procedure of Barnett and Seligman[180], as modified by Esterbauer[181,182]. The coloration is due to an absorption maximum at 18000 cm^{-1} (556 mm); the single cell is measured at this wavenumber by the scanning procedure, and from this the total extinction ($E_{tot.}$) is determined. Table 3.20 gives the results as found by Esterbauer et al.[183].

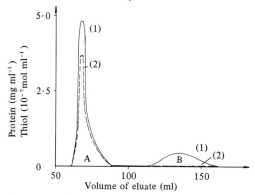

Figure 3.8. Separation of soluble cell components of Nemeth-Kellner lymphosarcoma into high molecular weight (A) and a low molecular weight (B) fractions, on Sephadex G25 (fine) (according to Rindler and Schauenstein[179]). (1) Sulphydryl concentration in the eluate; (2) protein concentration in the eluate.

Table 3.19. PSH_s and NPSH contents of EATC, and the influence on these of HPE (according to Rindler and Schauenstein[179]).

	PSH_s (mol of SH/mg of protein)	$-\Delta PSH_s$ (%)	NPSH (mol of SH/10^6 cells)
Controls	1·26 (±0·02) × 10^{-7}		1·39 (±0·14) × 10^{-9}
+HPE	0·64 (±0·08) × 10^{-7}	49	0

The values given in table 3.20 demonstrate a definite, strong decrease of $PSH_{tot.}$ after the HPE incubation. The main decrease in the $\bar{E}_{tot.}$ values, over the controls, is approximately 33%. A quantitative evaluation per mol of $PSH_{tot.}$ was not possible according to the original instruction of Barnett and Seligman[180], but Esterbauer[181,182] succeeded in finding the conditions for a quantitative reaction between the DDD reagent and the $PSH_{tot.}$. He also determined the molar extinction coefficient of the colour–protein complex formed between the protein and the azo dye. On the basis of these investigations an evaluation of the $\bar{E}_{tot.}$ values per mol of $PSH_{tot.}$ proved possible.

The strong decrease of $PSH_{tot.}$ after HPE treatment suggests, after what has been said about PSH, that a corresponding equivalent of HPE is bound to the proteins of the treated cells. This hypothesis was confirmed in the following way. To HPE-treated EATC, washed in isotonic sodium chloride solution, was added DNPH solution, followed by the removal of excess of reagent; spectrophotometric investigation of the cells was then carried out. Such cells already appear yellow on subjective microscopic observation, and exhibit in the microspectrophotometer the well-known absorption spectrum of the DNP-hydrazone of a saturated aldehyde, with a maximum at 27 250 cm^{-1} (367 nm) and a minimum at 32 250 cm^{-1} (310 nm). This is completely analogous with the spectra of proteins preincubated with HPE and subsequently treated with DNPH (cf page 47, and figure 3.6). Figure 3.9 shows such a spectrum[185].

As a control, EATC were preincubated in 0·9% sodium chloride solution alone, and on subsequent treatment with DNPH they likewise showed a measurable absorption at 27 250 cm^{-1}. The $\bar{E}_{tot.}$ values at this wavenumber, measured for over one hundred single cells, are given in table 3.21. The extinction, which is definitely conditioned by the HPE treatment, is thus given by the difference in $\bar{E}_{tot.}$ values. By considering the confidence limits of the mean $E_{tot.}$ values, one obtains a maximum value of 14·7 and a minimum value of 10·8. From this may be calculated the amount of cell-bound hydrazone, namely of cell-bound HPE, if the extinction coefficient for the DNP-hydrazone of HPE is known. The number

Table 3.20. Influence of HPE on the content of structural protein sulphydryl groups ($PSH_{tot.}$) in EATC (according to Esterbauer et al.[183]). $\bar{E}_{tot.}$ is the mean value of the total extinction of the single cell, determined for about 50 cells in each case. $E_{tot.} = \int E df$, where E is the extinction; df is as small an orifice as possible, usually ≤ 1 μm^2.

$PSH_{tot.}$	$\bar{E}_{tot.}$	$PSH_{tot.}$ (mol/cell)	Decrease in $PSH_{tot.}$ (mol/cell)
Controls	32·3	1·7 × 10^{-14}	0·54 (±0·18) × 10^{-14}
+HPE	22·0	1·16 × 10^{-14}	

of moles of hydrazone per single cell is equal to $\bar{E}_{tot.}/\epsilon$, or grams of hydrazone per single cell is equal to $\bar{E}_{tot.}/a$ where a is the specific extinction coefficient. These values of the extinction coefficient were found to be[159]: $\epsilon = 21\,000$ litre mol^{-1} cm^{-1}, or $a = 75$ litre g^{-1} cm^{-1}. Thus the quantities of HPE actually bound to the sulphydryl groups of the proteins are found to be between 0.5×10^{-14} and 0.70×10^{-14} mol cell^{-1}. This value agrees with the corresponding loss of PSH$_{tot.}$ (table 3.20), and provides proof that HPE has been chemically bound to the sulphydryl groups of the proteins of the cell.

Esterbauer[182] determined the content of trichloroacetic acid-insoluble proteins for EATC to be 170–190 μg 10^{-6} cells. From these values it follows that three molecules of HPE are bound per 10^5 g of protein. This constitutes a highly specific binding.

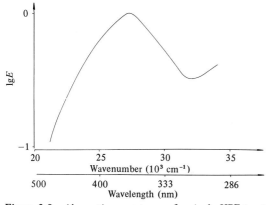

Figure 3.9. Absorption spectrum of a single HPE-treated Ehrlich ascites tumour cell, after staining with 2,4-dinitrophenylhydrazine, measured with the Zeiss instrument UMSP-I, objective UF 100, projective 100 (according to Burkl et al.[185]).

Table 3.21. Total extinctions of EATC treated with HPE and incubated with 0·9% sodium chloride solution, after treatment with DNPH (according to Scheuringer[184] or from measurements obtained in collaboration with Nöhammer[184a]).

	$\bar{E}_{tot.}$	Standard deviation	95% confidence limits
Controls	3·9	1·7	3·6–4·3
+HPE	16·7	9·9	14·9–18·4

3.2.6 Differences in effects of hydroxyenals (HE) on different processes in the same cell type, or on the same processes in different cell types

The previous sections have been concerned with the inhibitory effects of HE, in particular on tumour cells, having once shown that significant differences exist with tumour cells. The actions of HE on a broader spectrum of cell types will now be considered.

All the biochemical and biological activities of HE so far known are based on a reaction with biochemically or biologically functional sulphydryl groups. Hence it is natural to begin with the discussion of this reaction. As already mentioned, sulphydryl groups may possess very different reactivities towards HE. Thus it was pointed out that lower molecular weight thiol compounds react more rapidly than protein-bound thiol compounds (page 49), that different sulphydryl groups within one protein molecule, as well as of different proteins, exhibit very different reaction velocities (page 49).

These results referred to pure solutions rather than the polyphase conditions that occur within cells. In what follows it will be shown that sulphydryl groups within the cell exhibit very diverse reactivities. In general we are dealing here with nonprotein thiol compounds, that is, the sulphydryl groups of soluble and structure-bound cell proteins.

In particular we shall describe the different activities of sulphydryl groups of the various enzymes of one type of cell, as well as a specific enzyme system in different cell types. From these data the diverse biochemical and biological reactions of different types of cell towards HE will become comprehensible. Let us begin with the problem of the losses in sulphydryl groups suffered by different cell types in soluble proteins (PSH_s) and nonprotein thiol compounds (NPSH) under the influence of HPE.

Rindler and Schauenstein[179] and Schindler[186] have described the following findings.
(1) Normal rat liver shows an extremely high content of NPSH and PSH_s whereas normal rat kidney, jejunum, and diaphragm, as well as several animal tumours, have values of NPSH which are smaller by one to two orders of magnitude, and values for PSH_s are even smaller, compared with those for liver. This is shown in table 3.22. The values quoted as controls for the HPE incubations which will be discussed below were determined after 30 minutes incubation in isotonic sodium chloride solution at 37°C. Since during the incubation soluble cell components diffuse into the medium these cells are not in their original condition. The original NPSH values are higher by approximately 100%, those for PSH_s are higher by about 20–50%[53].
(2) After 30 minutes action of 5 mM HPE solution a drastic decrease of all sulphydryl values occurs:
(a) The NPSH values for kidney, jejunum, and diaphragm, and of all tumours investigated, drop to zero, whereas that of liver drops to a residual value of 20% of the controls. It could also be ascertained that this incomplete reaction of the NPSH of liver was certainly not connected with an insufficient concentration of HPE in the incubation medium. During the incubation the HPE concentration in the medium decreased only from 5 mM to 4·8 mM[186].
(b) The PSH_s values show appreciably smaller decreases, for which certain differences between different cell types can be recognized. Again the

Table 3.22. NPSH and PSH$_s$ content of several normal organs (measured in tissue slices) of the rat, and of several transplantable tumours of the mouse in ascites fluid and solid forms, after 30 minutes incubation in 0·9% sodium chloride solution. The values are given in μmol of thiol groups/g wet weight.

Preparation	NPSH	PSH$_s$
Rat liver	2·5-3·0	5·8
Rat kidney	0·5	3·4
Rat jejunum	0·06	0·8
Rat diaphragm	0·09	1·7
EATC (Heidelberg)	0·5	3·8
EATC (London)	0·6	–
Sarcoma-180 (ascites)	0·8	2·8
Nemeth-Kellner lymphosarcoma (ascites)	0·7	2·6
Ehrlich solid tumour (Heidelberg)	0·37	1·4
Sarcoma-180 (solid)	0·27	1·6
Nemeth-Kellner lymphosarcoma (solid)	0·24	1·25

Table 3.23. Action of HPE (5 mM) on the PSH$_s$, referred to 1 mg of the soluble cell proteins, of several normal organs of the rat and transplanted tumours (according to Schindler[186]).

Tissue	PSH (10^{-7} mol/mg protein)		$-\Delta$PSH$_s$ (10^{-7} mol/mg protein)	Inhibition of oxygen uptake in the first 10 min, after 30 min preincubation in 5 mM HPE$_s$ (%)
	30 min, 0·9 % NaCl	30 min, 5 mM HPE$_s$		
Rat liver, normal	1·47 ± 0·02	1·45 ± 0·04	0·02	0
Rat kidney, normal	0·79 ± 0·04	0·63 ± 0·04	0·16	17
Rat jejunum, normal	0·57 ± 0·04	0·34 ± 0·05	0·23	27
Rat diaphragm, normal	0·88 ± 0·09	0·66 ± 0·07	0·22	31
Rat DENA hepatoma stage 1[186]	1·03 ± 0·02	0·97 ± 0·04	0·06	0
stage 2[186]	0·84 ± 0·06	0·84 ± 0·04	0	10
Rat Yoshida sarcoma	1·16 ± 0·05	0·84 ± 0·04	0·18	50
EATC (Heidelberg)	1·26 ± 0·02	0·64 ± 0·08	0·62	100
EATC (London)	1·50 ± 0·04	0·89 ± 0·04	0·61	100
Sarcoma-180	1·27 ± 0·06	0·76 ± 0·06	0·51	100
Nemeth-Kellner lymphosarcoma	1·21 ± 0·03	0·71 ± 0·02	0·50	100
Ehrlich solid tumour (Heidelberg)	0·71 ± 0·09	0·47 ± 0·09	0·24	~58
Nemeth-Kellner lymphosarcoma (solid)	0·90 ± 0·05	0·72 ± 0·06	0·18	~41
Sarcoma-180 (solid)	1·18 ± 0·12	0·83 ± 0·10	0·27	~28

unique position of the liver is made apparent, its PSH_s value remaining practically unchanged. The remaining healthy organs studied, the Yoshida sarcoma, and the solid forms of the Ehrlich carcinoma, sarcoma-180, and lymphosarcoma, exhibit equally large sulphydryl losses of the soluble proteins, whereas four ascites tumours show two to three times higher PSH_s losses. A special position is taken up by the diethylnitrosamine (DENA) hepatoma, which, like healthy rat liver, suffers practically no PSH_s loss (table 3.23). The remarkable finding that, with all the cell types investigated here, the NPSH decreases under the influence of HPE appreciably more strongly than does the PSH_s, was examined in a somewhat more detailed manner for EATC. In figure 3.10 the ordinate shows the NPSH and PSH_s losses as a percentage of the corresponding controls; the abscissa presents the molar proportions for the different HPE concentrations in the medium, namely mol of HPE/mol of total thiol groups. By means of addition of PSH_s ($11 \cdot 2 \times 10^{-9}$), PSH_i ($3 \cdot 8 \times 10^{-9}$) and NPSH ($2 \cdot 7 \times 10^{-9}$) the total sulphydryl content of the cells was calculated to be about $17 \cdot 7 \times 10^{-9}$ mol/10^6 cells[186].

Tables 3.24 and 3.25 present the results together with the significance limits. With 0·4 mM HPE the PSH_s do not yet show a statistically significant decrease, and 93% of the NPSH have already disappeared. It is seen that with increasing HPE concentration the NPSH are first used up, whereas the PSH_s have not yet been attacked. The latter enter into

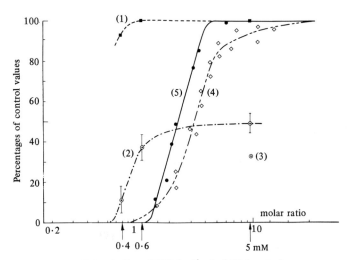

Figure 3.10. Decrease in NPSH, PSH_s and PSH_i, or inhibition of respiration and glycolysis, effected by HPE in Ehrlich ascites tumour cells, as percentages of control values. Values are dependent on the appropriate molar ratio HPE/total sulphydryl groups of the cells (according to Schindler[186]). Arrows indicate corresponding mM concentrations of HPE in the incubation medium. (1) NPSH; (2) PSH_s; (3) PSH_i; (4) inhibition of respiration; (5) inhibition of glycolysis.

reaction only when small amounts of NPSH remain. The graphical presentation of figure 3.10 suggests that a stepwise reaction takes place, first with the NPSH, and then with the PSH_s of the cell. In this way NPSH take over a protective function for the sulphydryl groups of the soluble proteins, and obviously also for those of the trichloroacetic acid-insoluble proteins. This is shown in figure 3.10 by the encircled point (⦿), which corresponds to the percentage decrease of structural protein sulphydryl groups, determined cytospectrometrically[183]. The preferred reaction of the NPSH may be explained by their greater accessibility to HPE, whereas the PSH_s are partly more difficult of access or not accessible at all (masked sulphydryl groups), that is to say they remain less reactive, because of the structure of the proteins and the cell compartments. In addition there is an enzymatic factor, namely the GSH-S-alkene transferases

Table 3.24. NPSH and PSH_s values of EATC after incubation (30 min, 37°C) in various HPE concentrations (according to Schindler[186]). n is the number of separate experiments.

n	10^{-9} mol of SH groups/ 10^6 cells		mg protein/ g wet weight	mol SH groups/ mg protein	Incubation medium
	NPSH	PSH_s			
8	$1 \cdot 39 \pm 0 \cdot 14$	$9 \cdot 70 \pm 0 \cdot 60$	$26 \cdot 5 \pm 1 \cdot 7$	$0 \cdot 126 \pm 0 \cdot 002$	0·9% NaCl solution
3	$0 \cdot 00$	$4 \cdot 29 \pm 2 \cdot 32$	$19 \cdot 4 \pm 5 \cdot 5$	$0 \cdot 064 \pm 0 \cdot 008$	5 mM HPE
4	$1 \cdot 26 \pm 0 \cdot 24$	$7 \cdot 13 \pm 1 \cdot 32$	$24 \cdot 4 \pm 5 \cdot 4$	$0 \cdot 103 \pm 0 \cdot 007$	0·9% NaCl solution
4	$0 \cdot 00$	$4 \cdot 08 \pm 0 \cdot 79$	$21 \cdot 5 \pm 3 \cdot 9$	$0 \cdot 065 \pm 0 \cdot 007$	0·6 mM HPE
5	$1 \cdot 85 \pm 0 \cdot 23$	$8 \cdot 16 \pm 0 \cdot 78$	$23 \cdot 5 \pm 3 \cdot 2$	$0 \cdot 104 \pm 0 \cdot 008$	0·9% NaCl solution
5	$0 \cdot 19 \pm 0 \cdot 07$	$6 \cdot 67 \pm 1 \cdot 87$	$20 \cdot 4 \pm 4 \cdot 5$	$0 \cdot 092 \pm 0 \cdot 007$	0·4 mM HPE

Table 3.25. Decrease of PSH_s in EATC after incubation in various concentrations of HPE. In the last column results are given of Student's t-test: limits of significance $2\alpha^a = 0 \cdot 05$ (according to Schindler[186]).

Concentration of HPE (mM)	Molar ratio HPE/$SH_{tot.}$ (approx.)	$-\Delta PSH_s$/mg protein	Percentage of controls	Significance
5	9·88:1	$0 \cdot 062 \pm 0 \cdot 006$	49	$2P^b < 0 \cdot 001$; max significance
0·6	1·19:1	$0 \cdot 038 \pm 0 \cdot 007$	37	$2P < 0 \cdot 01$; highly significant
0·4	0·79:1	$0 \cdot 012 \pm 0 \cdot 007$	11·5	$0 \cdot 3 > 2P > 0 \cdot 2$; not significant

[a] α is the postulated probability of significance.
[b] P is the actual probability of significance (cf *Documenta Geigy*, 7th edition, 1969, page 150).

discovered by Boyland *et al.*, and described in detail in section 3.3. These enzymes specifically catalyse the coupling of GSH with α,β-unsaturated carbonyl compounds. According to these very interesting results of Boyland *et al.* it is the liver in particular that exhibits the highest activities of such transferases, kidney and other organs exhibiting only a fraction thereof.

It appears therefore that liver and DENA hepatoma exhibit no measurable loss of PSH_s after treatment with a 5 mM HPE solution, because of their excessively high NPSH contents[186,187] and the high transferase activities. The presence of the corresponding differences in the activities of GSH-S-alkene transferases in the different organs and tumours may explain the very different PSH_s losses, after administration of HPE, incurred by kidney, jejunum, and diaphragm, and by the tumours named in table 3.22, although they possess approximately equally high NPSH values. Figure 3.11 shows for the twelve animal tumours an unambiguous connection between PSH_s loss and proliferative activity. The latter is expressed as the doubling time, obtained from the corresponding growth curves or the increase in cell counts[186]. For DENA hepatoma the tumour cell growth was calculated from the increase of the liver weight, according to Heise and Görlich[188,189]. The semi-logarithmic plot in figure 3.11 shows that, with tumours with doubling times between about 2 and 7 days, apparently an exponential function exists between the number

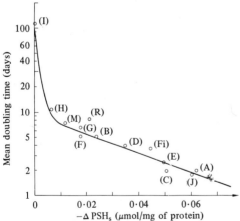

Figure 3.11. Connection between PSH_s loss caused by 5 mM HPE, and the doubling time of various transplanted tumours in animals (according to Schindler[186]).
(A) EATC (Heidelberg); (B) Ehrlich solid tumour (Heidelberg); (C) sarcoma-180, ascites form; (D) sarcoma-180, solid; (E) Nemeth-Kellner lymphosarcoma, ascites; (F) Nemeth-Kellner lymphosarcoma, solid; (G) Yoshida sarcoma; (H) diethylnitrosamine hepatoma, stage 1; (I) diethylnitrosamine hepatoma, stage 2; (J) EATC (London); (Fi) chicken fibroblasts in culture; (R) rhabdomyosarcoma; (M) H.P.-melanoma[186a,b,c].

of HPE-blocked PSH_s/mg of soluble proteins and the dividing activity:

$$\lg t_D = A - B(\Delta PSH_s) .$$

Hence we may conclude that there exist specific PSH_s which are responsible for cell division and exhibit highly selective sensitivity to HPE. It seems probable that these thiol compounds belong to the acidic residual proteins bound to chromosomal DNA, as well as to the DNA-polymerase itself. As table 3.23 shows, PSH_s damage from the effect of HPE *in vitro* also occurs in the tissues of healthy organs. Since we are not concerned here with an actual increase in cells, but with the substitution of newly-formed cells for dead ones, the use of the concept of doubling time seems meaningless. The values of ΔPSH_s lie between $0 \cdot 002$ and $0 \cdot 02$ μmol/mg protein. Probably, the range of validity of the above exponential rule also ends here. As the value for normal chicken fibroblasts in culture shows, see figure 3.10, the rule seems to hold not only for malignant cells, but for rapidly dividing cells in general ($t_D < 7$ days).

It therefore seems informative to study those biochemical or biological functions of the cell which are blocked by HPE through reactions with specific functional PSH_s, and to investigate if a more precise criterion may emerge.

Let us first consider cell respiration, which as is well known involves a multitude of sulphydryl proteins. From table 3.23 it can be seen that inhibition of respiration by HPE differentiates a little more clearly between tumour cells and normal cells.

Bickis *et al.*[165] obtained essentially the same results. With EATC, $0 \cdot 45$ mM HPE caused a 50% inhibition of respiration, whereas with normal human leucocytes $4 \cdot 5$ mM HPE was required to produce an equally strong inhibition. The experimental procedure was somewhat altered: instead of preincubation with HPE, the aldehyde was added to the cells suspended in glucose–phosphate buffer at the beginning of the one-hour experiments. For this, $0 \cdot 1$ mM HPE was required for EATC and sarcoma-37 cells to produce a 93%–97% inhibition of respiration, but with rat jejunum, spleen, and thymus, a $0 \cdot 4$ mM HPE solution produced 83%–94% inhibition.

Analogous results obtained with HOE for EATC and normal rat liver and rat kidney have already been referred to (see section 3.2.2).

Further cell functions which are inhibited by HPE, as a result of reaction with functional enzyme sulphydryl groups, are those of the biosyntheses of nucleic acids and proteins. Bickis *et al.*[165] have also found here very strongly marked differences between tumour cells and normal tissues, and these are presented in table 3.26. Susceptibility of the biosyntheses to HPE shows the same sequence, namely DNA > protein > RNA, for all the preparations investigated. For normal tissues, however, at least four times the HPE dosage is required to produce the same extent of inhibition.

So far we have mentioned only the different reactions which HPE undergoes with the soluble NPSH and PSH_s, and the biochemical blockings of metabolism which they cause. The reaction of HPE with the sulphydryl groups of insoluble structural proteins of the cell (PSH_i), and the blocking of specific metabolic steps which therefore arise, also show marked differences with various cell types.

The inhibition of succinate oxidation was examined in more detail, to provide our first example. As already mentioned, this process is decisively involved in the inhibition of cell respiration caused by HPE[173]. The parameters measured, as mentioned on page 58, were the enzymatically determined succinate content, and the intensity of coloration with the MTT reagent[190], measured cytospectrometrically, on succinate dehydrogenase activity. The results thus obtained are presented in tables 3.27 and 3.28. Details of the experimental technique, together with the data for EATC, are given by Schauenstein and Höfler-Bergthaler[190]. The remaining values of table 3.28 are taken from unpublished measurements obtained in collaboration with E. Höfler-Bergthaler.

Both series of measurements show a highly significant inhibition of succinate metabolism in five tumours after action of HPE. A comparison of the percentage accumulation of succinate with the percentage decrease in SDH activity shows that deactivation of SDH is crucially involved in the inhibition. The preferential action of HPE on malignant cells becomes apparent in both series of measurements; kidney and, again, liver show the highest resistance towards HPE with regard to the structure-bound PSH_s in their SDH.

Table 3.26. Comparison of the action of HPE *in vitro* on the respiration and biosyntheses in ascites tumour cells and several organs (measured in tissue slices) of the rat (according to Bickis *et al.*[165]). Incubation was for 1 h in Krebs–Ringer phosphate glucose medium.

Preparation (and concentration of HPE)	Inhibition of oxygen uptake (% of controls)	Incorporation (% of controls)		
		leucine into proteins	adenine into RNA	adenine into DNA
EATC (0·1 mM)	97	70	90	40
Sarcoma-37 ascites cells (0·1 mM)	95	72	90	48
Slices of rat jejunum (0·4 mM)	93	96	110	81
Slices of rat spleen (0·4 mM)	83	77	78	58
Slices of rat thymus (0·4 mM)	94	71	83	56

α,β-Unsaturated aldehydes 73

It is also interesting to compare the SDH activities in the control experiments (table 3.28). These show that liver, as might be expected, possesses by far the highest SDH activity, followed by the cells of healthy monkey kidney cortex, whereas all the four tumours examined show far lower values. In view of these marked differences also in the action of HPE on the PSH_i, one is tempted to search for the solution to the problem.

Table 3.27. Accumulation of succinate in tumours and normal tissue slices, after 30 minutes incubation at 37°C in HPE, expressed as percentage of control values (incubation in 0·9% sodium chloride solution) (according to Schauenstein and Kapfer[173] or Kapfer and Schauenstein[191]).

Preparation	Concentration of HPE in the incubation medium		
	3 mM	5 mM	10 mM
EATC	95·0		
Nemeth-Kellner lymphosarcoma (ascites)	45·0		
Sarcoma-180 (ascites)	114·0		
Harding-Passey melanoma (solid)	59·5		
Rat liver (normal)	0		0
Rat kidney (normal)	0	31·4	49·3
Rat spleen (normal)	20·7	38·8	72·2

Table 3.28. Activities of succinate dehydrogenase in tumours and normal tissues after incubation (20 minutes at 37°C) in HPE (3 mM), expressed as the intensity of colour formation with MTT reagent (according to Schauenstein and Höfler-Bergthaler[190]. Values shown are the mean total extinctions ($\bar{E}_{tot.}$) for each of two series, (a) and (b), consisting of 40-100 single cells (dispersed cells from slices of kidney and liver). In order to test the SDH-specificity of the staining, experiments were carried out with EATC treated with $CoCl_2$ MTT-reaction mixture lacking succinate. Under these conditions practically no extinction was measured.

Preparation	Series	NaCl	HPE	Mean change (%)
Sarcoma-180	(a)	27·0	16·2	−50
	(b)	20·6	7·7	
EATC [190]	(a)	32·0	7·0	−79
	(b)	20·6	4·4	
Nemeth-Kellner lymphosarcoma	(a)	40·3	11·7	−74
	(b)	25·5	5·6	
Plasmocytoma G		16·5	3·9	−76
Monkey kidney cortical cells	(a)	71·1	85·8	+23
	(b)	38·4	48·3	
Rat liver cells	(a)	322·0	325·6	+1

A first experiment in this direction was undertaken by Zollner and Schauenstein[192] with EATC and normal rat liver. With mitochondria isolated from untreated EATC and liver cells, it was found that the succinate dehydrogenase of both cell types possessed the same sensitivity towards HPE. Half-maximal inhibition of ADP-stimulated succinate respiration is achieved by 5–7 μmol litre^{-1} mg^{-1} of protein of liver mitochondria and 6 μmol litre^{-1} mg^{-1} of protein of EATC mitochondria. If the reduction of MTT reagent (see pages 58, 72, 73) in the presence of succinate is examined, that is, the activity of the succinate–MTT oxidoreductase, one observes a similar inhibition by HPE, which is virtually equally great for the mitochondria of both cell types. If, however, whole EATC or liver cells are incubated in HPE (1 mM) for 30 minutes at 37°C, and then the mitochondria are isolated, their succinate respiration can then be examined. For liver mitochondria an insignificant 5% inhibition is found, compared with controls, but with EATC mitochondria there is 75·3% inhibition. The activity of succinate–MTT oxidoreductase in EATC incubated with HPE is inhibited half-maximally by 0·8 mM HPE, but with whole liver cells incubated in HPE only by 5·4 mM HPE (table 3.29). From these results one may conclude that the different susceptibilities towards inhibition of the SDH in the liver cells and EATC are not caused by differences in sensitivity to HPE, but by extramitochondrial factors. As such we may consider:

(1) the extremely high content of NPSH, whose protective function for protein sulphydryl groups has been discussed above; (cf page 68)

(2) enzymes, which metabolize HPE to conversion products that are no longer inhibitory, especially GSH-S-alkene transferases, which have been found to be present in high activity in cytoplasm (cf section 3.3). They may also possibly be oxidoreductases specific for α,β-unsaturated carbonyl compounds;

Table 3.29. Influence of HPE on succinate–MTT oxidoreductase in $2·86 \times 10^6$ liver cells and EATC corresponding to 1·49 and 0·63 mg of protein respectively. Values in column 3 are the fall of extinction per 15 minutes.

Cell type	HPE (mM)	Succinate–MTT oxidoreductase activity	Inhibition (%)
Liver	0	1·610	0
	3	1·080	33
	10	0·282	83
	30	0·024	99
EATC	0	0·191	0
	0·5	0·151	21
	1	0·077	60
	3	0·025	97

(3) the appreciably higher content of mitochondria. It may be seen from table 3.29 that the SDH activities, based on mg of protein, of tumour cells are a quarter those of liver cells, whereas the isolated mitochondria of both cell types possess practically the same SDH activity. A lower content of mitochondria in the cells signifies a lower content of SDH, whose reactive sulphydryl groups are consequently present in lower molar proportions in relation to the proffered HPE. These groups therefore suffer a greater probability of attack than in a cell with a higher mitochondrial content. Also the cytospectrometrically determined differences in SDH activities of liver and kidney, on the one hand, and the four tumours, on the other, as given in column 3 of table 3.28 are probably based on differences in content of mitochondria. Hence the difference in mitochondria content may explain the differences in inhibition by HPE in normal cells and in tumour cells. It was thus shown that an extremely high NPSH content does not represent the only active protective factor of the PSH_i towards HPE in the normal cell, but that the molar ratio of particular PSH_i/HPE could be decisive here, for example in kidney, which contains appreciably more PSH_i (of the SDH) than do the four tumours used for comparison.

3.2.7 Biological consequences

From the numerous and varied biochemical effects of hydroxyenals in general, and of HPE in particular, we can also anticipate corresponding biological effects.

Scaife[193] has examined the effect of HPE on mitosis in human kidney T-cells in synchronous culture. Figure 3.12 shows the results of these interesting experiments. Scaife concluded that HPE is about seven times as active an inhibitor as is kethoxal. Table 3.30 shows the differences in cytotoxic effects of HPE on rapidly proliferating cells, and on those which are blocked in the S-phase by thymidine in excess or are contact-inhibited by the corresponding cell density. Also of interest is the observation by Scaife that nonmitotic small thymocytes of the rat are hardly attacked by cytotoxic doses of HPE, despite the fact that they show rapidly degenerative nuclear pyknoses with low concentrations of many cytotoxically active substances. Human HeLa carcinoma cells were as sensitive as nonmalignant human kidney T-cells. Scaife summarized his results by stating that HPE is a cytotoxically active compound which reveals its activity most readily with actively proliferating cells, and rather less so with nondividing cells. It remains to be shown whether HPE exhibits differences in its cytotoxic effect between proliferating normal cells and malignant cells *in vivo*.

The investigations to be discussed now were concerned with a problem which is fundamental to the study of hydroxyenals in general. The consequence of the high affinity of HE for sulphydryl groups is an extraordinarily short persistence within the organism. On intravenous injection into mice or rats, after only a few minutes the amounts of HE in the blood are undetectable. When HE is injected intraperitoneally only

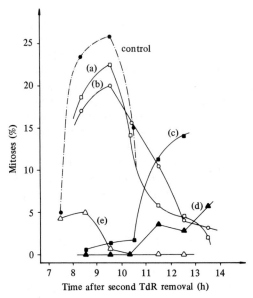

Figure 3.12. Effect of HPE on mitosis in synchronized cultures of human kidney T-cells (according to Scaife[193]).
(a)–(d) HPE was added in S-phase (2·5 h after removal of the second Td-R block):
(a) 5 μM; (b) 10 μM; (c) 50 μM; (d) 100 μM.
(e) HPE (100 μM) was added in G_2-phase (7 h after removal of the second Td-R block).

Table 3.30. Cytotoxic effect of 4-hydroxypentenal on mammalian cells under different conditions of proliferation (according to Scaife[193]).

Cells	Concentration of HPE			
	10 μM	50 μM	100 μM	500 μM
Kidney T-cells:				
proliferating,	−	+	+++	++++
blocked in S-phase	−	−	+	++
3T3 cells:				
proliferating,	−	+	+++	++++
contact-inhibited	−	−	+	++
HeLa cells:				
proliferating	−	+	+++	++++
Rat small thymocytes				
(% pyknosis)	11	11	13	14

Explanation of symbols: − indicates no visible damage; + indicates visible damage; increasing numbers of +s indicate increasing extent of damage.

traces of it can be found in the peritoneal cavity two to three minutes later, and in venous blood virtually none is found. One should not expect HE therefore to influence a growing implanted tumour on intravenous or intraperitoneal injection—yet Burk and Woods[162], using such a method of application of HOE, found a selective action on the melanoma S91 and the Ehrlich ascites tumour, as reported in section 3.2.2. The observed effects represent only the inhibition of the energy metabolism, however, and not a regression of the tumours, so that one cannot yet claim a therapeutic effect, as has actually been found by Conroy et al.[162a]. In the subcutaneous region of the skin of the back, which is less well supplied with blood, the situation is more favourable for a persistence of the hydroxyenals. The mean half-life based on measurements carried out in collaboration with Wünschmann[163] was found to be about two to three minutes.

Initial experiments were conducted with homologous five- to eight-carbon atom hydroxyenals. Injection of HPE around a subcutaneous implant of 10^7 EATC effectively destroyed it. Hydroxyoctenal was equally effective, but since HPE has a higher water solubility (cf table 3.12) therapeutic experiments with HPE were undertaken only with the solid forms of the Ehrlich tumour[194], Nemeth–Kellner lymphosarcoma[196], sarcoma-180[195], and Harding–Passey melanoma[197]. The procedure adopted for this was fundamentally as follows. The tumour cells were harvested under sterile conditions from the ascites fluid of mice, namely cells of Ehrlich ascites (strain Heidelberg), sarcoma-180, and Nemeth–Kellner lymphosarcoma, on the seventh to the tenth days *post transplantationem*. Thereupon 10^7 cells in 0·1 ml of ascites fluid were implanted under the skin of the back of each animal (IR-albino mice of both sexes, 20–25 g). With Harding–Passey melanoma 6×10^6 melanoma cells per animal (NMRI mice, female, 20–25 g) were implanted in the right flank.

The therapy was commenced with Ehrlich tumour, sarcoma-180, and Nemeth–Kellner lymphosarcoma on the third and the eighth days, and with melanoma seven weeks, *post implantationem*. The therapy for the three day old Ehrlich tumour consisted of a double dose, each of 2×10^{-7} mol of HPE/g body weight, given with an interval of 15 minutes on the third day *post implantationem*. With the eight day old tumour a double dose was given as above on the ninth and twelfth days *post implantationem*. With the three day or eight day old sarcoma-180, and the Nemeth–Kellner lymphosarcoma, a double dose, each of 2×10^{-7} mol, was again delivered either on the fourth and the seventh days or, alternatively, on the ninth and the twelfth days *post implantationem*. With melanoma either a single dose, of 2×10^{-7} mol/g body weight, or two separate doses (as above) with an interval of 15 minutes, or three separate doses with intervals of 15 minutes were given. All single doses were applied in 1 ml solutions, peritumorally and subcutaneously, with careful avoidance of injury to the tumour. The results of the therapy are summarized in table 3.31.

The therapeutic doses in all the experiments were without exception well tolerated by healthy animals. With approximately 80% of the experimental animals, local skin damage occurred at the place of injection, with a diameter of 0·5-1 cm. This damage could also be observed with some 20% of the matched controls but occurred there more weakly. These skin defects healed with the formation of a smooth scar, but after a period of twenty-one days no fur regeneration could be seen. Losses of the experimental animals amounted in all to some 10%-12%, and with the matched controls throughout, 0%. In the column headed 'Delayed' in table 3.31, for the first three tumours mentioned only those animals were considered whose tumour weight lay between zero and the weight of the control tumours, reduced by double the error of the mean value. For the tumours designated 'Delayed' the number of strongly inhibited (75%-100%) tumours far exceeded the others; their numbers amounted to between 25% and 58%. Those less strongly inhibited (50%-75%) were found in 10%-20% of cases, and those weakly inhibited in 3%-15% of cases. With the melanoma the strongly ($\geqslant$83%) and rather less strongly (~50%) inhibited tumours represented approximately 20% each of the totals.

The activity indices cited are the quotients of the tumour weights of the treated animals and of the control animals. The controls received injections of isotonic sodium chloride solution. According to an international agreement a quotient of 0·0-0·3 signifies a marked or good tumour inhibition, of 0·4-0·6 a moderate to weak inhibition, and 0·7-1·0 no tumour inhibition (representing the scatter about the controls).

Table 3.31. Results of therapy with HPE on transplantable tumours of animals.

Tumour	Prevented (%)	Delayed (%)	No effect (%)	Activity index [a]	Deaths of animals (%)
Ehrlich[194]					
3 days	31	49·6	19·4	0·23	10
8 days	35	61	4	0·23	12
Sarcoma-180 [195]					
3 days	30	59	11	0·19	5
8 days	8	73	19	0·38	6
Nemeth-Kellner lymphosarcoma [196]					
3 days	18	76	6	0·17	4
8 days	2	68	30	0·54	6
Melanoma[197]					
7 weeks (0·2 M HPE)	8	66	25	0·37	0
7 weeks (0·4 M HPE)	40	47	0	0·12	13
7 weeks (0·6 M HPE)	40	33	0	0·09	27

[a] Tumour weight of treated animals/tumour weight of control animals (see the text).

Table 3.31 therefore shows that HPE, on subcutaneous peritumoral application, definitely inhibits the growth of the four tumours named (see plate 1). The inhibitory action is appreciable and good with three- and eight-day old Ehrlich tumours, with three-day old sarcoma-180, and with two-day old lymphosarcoma. With seven-week old melanoma the inhibition is designated as very good. The eight-day old sarcoma-180 and lymphosarcoma responded only moderately. The appreciably better inhibition found for three-day old sarcoma-180 and three-day old lymphosarcoma, compared with the three-day old Ehrlich solid tumour, may be due to the double doses of HPE that were used for these tumours. The decrease in activity with the eight-day old sarcoma-180, which does not occur with the Ehrlich tumour, is an expression of the well-known high resistance of sarcoma-180 towards therapy. A similar situation may also obtain for lymphosarcoma. With this tumour a striking dependence was further shown, both for the control animals and the treated animals, of tumour growth on the sex of the animals. The mean weight of tumour in the controls was for male animals 1·5 times that for the female animals, as is shown in table 3.32. For the animals given therapy it was shown that in the males the more rapidly growing tumour is many times more strongly inhibited by HPE (activity index 0·12) than for the females (activity index 0·4). The mean tumour weight for the male animals used in the therapy experiments was only half that of the females. This difference, according to the unilateral Wilcoxon test, is significant for $P = 0·05$. These surprising and highly significant results again suggest that the rapidly proliferating cells possess a higher sensitivity to HPE than those proliferating more slowly.

A similar situation exists for other effects produced by HPE, namely loss of PSH_s (page 70, figure 3.11), and inhibitions of the DNA/protein/RNA synthesis system (page 71), of mitoses (pages 75, 76), of respiration (table 3.23), and of SDH activity (table 3.27). The selective action of HPE on tumour cells growing in healthy tissues is clearly observed in the animals that have been cured. With these smooth scar-free healing is always observed (table 3.15). These results suggested that it would be of interest to examine the effect of a local HPE application on humans. Musger[198-200] was the first to carry out such exploratory experiments with five basaliomata which had been verified clinically and histologically.

Table 3.32. Sex-dependent growth of lymphosarcoma (Nemeth–Kellner) of the mouse (according to Schauenstein et al.[196]).

Sex	Tumour weight ± f_M	Number	Differences significant for
Male	0·636 ± 0·09	25	$2P^a = 0·02$
Female	0·396 ± 0·05	25	

[a] P is actual probability of significance.

The application involved covering the cancer with linen cloths steeped in aqueous 5% HPE solution, and then wrapping in an airtight manner. The period of treatment lasted eighteen to twenty-three days, and the following results were obtained. In three cases the basaliomata had healed with a gentle scarring (plate 2). The period of further observation was ten months. In two instances, one four weeks and the other seven months after the termination of treatment, histologically ascertained recidivious tumour cells bordering the tumour remained, but could likewise be healed by the same treatment. It was established histologically that the depth of action of HPE sufficed for the procedure of application adopted.

Ratzenhofer *et al.* [201] undertook comprehensive investigations of the effects of HPE on normal and carcinomatous human cervix. For preference HPE was used in the form of an alginate gel, 5% and 3% with respect to HPE, in a pessary cap. It was shown that HPE used in these concentrations also caused severe tissue damage to normal mucous membranes, and this is followed by a serous to necrotizing inflammation of the mucous membranes. Tissue death affects the surface epithelia fairly evenly, as also the surface tissue glands, and the stroma, including blood vessels and cells. Significant differences in these aldehyde defects were not observed for normal flat and highly prismatic cervical epithelia, nor with cancerous flat epithelia. Repeated applications under favourable conditions led to a series of tissue necroses 4–9 mm in depth. In two out of eight cases carcinomata restricted to the cervical surface were wholly destroyed, as verified histologically *in situ*. For the destruction of the formations of cancerous cells which had grown deep into the glands, the depth of action of the aldehyde proved insufficient.

It can be readily understood that conditions in the cervix, so richly supplied with blood vessels, are incomparably less favourable for the selective carcinostatic action of HPE than, say, the skin of the back or flank of experimental animals[194-197], which is much less well supplied with blood, or the facial skin in man[198]. On the one hand the good blood supply of the cervix requires high concentrations of HPE, which is known to be very unstable in the blood *in vitro* and *in vivo*; on the other hand the mucous membranes of the cervix are very sensitive to HPE. The problem is to find the correct concentration, which does not yet cause substantial damage to healthy tissue but which may already destroy the carcinoma. The magnitude of the experimental effort needed for this is formidable and has proved forbidding. We thus decided on a high concentration of HPE, about one hundred times higher than that used for the animal experiments; this corresponds roughly to that used for human basaliomata. Clearly the actual effective concentrations were still so high that a drastic loss of selective action on the carcinoma cells resulted. Considering that the concentration range for selective action of HPE in this special instance is unknown, however, such a result is not unexpected.

From the above it becomes clear that for therapeutic experiments with HPE, one essential difficulty consists in supplying sufficient quantities of the substance to the tumour cells *in vivo*, because of the generally high affinity of HPE to SH-groups.

Recent experiments, however, showed that it was possible to cope with this problem by making use of both the ability of HPE (and of other α,β-unsaturated aldehydes) to form reversible adducts with cysteine and of the particularly high affinity of HPE to protein thiols, these being of functional importance for the division of faster growing tumour cells.

It has already been mentioned (see section 3.1.2.1) that Esterbauer *et al.* have carefully studied the mechanism of the reactions between α,β-unsaturated aldehydes and SH-compounds, and analyzed the products so formed. We shall mention here only the HPE-Cys 1:1 adduct (cf scheme 1.5 in the introduction).

The blockade of the functional group of HPE brings about severe alterations of the biochemical and biological properties[198a] as follows:
(a) Drastic reduction of the toxicity to mice. For a single dose applied intraperitoneally the LD_{50} is reached with 0·8 mg of free HPE/10 g body weight, whereas single doses of 40 mg and more of the adduct/10 g body weight gave no macroscopically visible toxic effects in mice.
(b) Inhibition of [^{3}H]thymidine incorporation into EATC *in vitro*. After 2 hours incubation *in vitro* the incorporation of [^{3}H]thymidine into DNA of EATC is inhibited completely by a 100 mM solution of the adduct, whereas free HPE causes the same effect in 0·1 mM solution.
(c) Complete inhibition of tumour growth after preincubation *in vitro*. EATC incubated for 2 hours in 10 mM solution of the adduct did not proliferate at all after reimplantation into healthy mice; corresponding controls showed a growth rate of 100%.

By means of cytospectrometry it was possible to show that HPE, having been bound reversibly to cysteine, is gradually released to the same extent as it is consumed by intracellular protein SH-groups, following in principle the reaction scheme 3.9, set out in section 3.2. This, of course, provides an explanation for the effects described in (b) and (c) immediately above.

As a consequence, therapeutic experiments were undertaken with mice bearing the Ehrlich ascites tumour. When the adduct (3·5 mg/10 g body weight) was given intraperitoneally 72, 96, 120, 144, and 168 hours *post implantationem*, four out of ten animals were macroscopically free both from ascites and solid tumours at the 90th day, one was free from ascites but had a solid tumour. Five animals died before the 90th day both with ascitic and solid tumours but nonetheless with a highly significant increase in survival time of 137% compared with the controls, all of which had both ascitic and solid tumours[198a].

Hence it seems that the reversible binding to certain SH-compounds, such as cysteine, appears to be a promising way of transforming HPE (and other potential tumour-affecting α,β-unsaturated aldehydes) into stable storage forms from which they can be released at the sites of their intended action.

At about the same time, Conroy et al.[198b] undertook some similar experiments, but with a preparation of the HPE-Cys 1:2 adduct. The authors were able to report distinct inhibitions of [^{3}H]thymidine incorporation into cells of sarcoma-180 of mice after incubation in vitro, and a highly significant therapeutic effect (prolongation of survival time by 129%, maximally, but no tumour-free animals) with a single dose of 2·56 mg/10 g body weight given 72, 96, 120, 144, and 168 hours post implantationem.

The LD_{50} of the preparation amounts to 12·3 mg/10 g body weight. In contrast to these results, we have found the highly purified 1:2-adduct to be therapeutically ineffective to EATC and nontoxic[198a].

In section 3.2.5.1 it was stated that incorporation of adenine into DNA[165], or thymidine into EATC[177], responds particularly sensitively to HPE, and that the inhibitory effect proceeds many times more strongly with EATC and sarcoma-37 cells than with the normal cells used for comparison.

Different types of cellular material were, of course, used for these investigations. Thus it could not be decided from the results whether DNA synthesis as such is less sensitive to HPE in normal cells than in tumour cells, or whether, as with succinate dehydrogenase, factors (perhaps NPSH) are present in normal cells which protect DNA synthesis from the action of HPE. One might be able to make a contribution to this biologically interesting question if two different DNA syntheses could be made to proceed in the same cell type. This is possible with cells infected by a virus. An investigation of this kind was first undertaken[202,203] with cultured chick fibroblasts, which had been infected with vaccinia virus, a DNA virus of the smallpox group. The parameters measured were the incorporation of [methyl-^{3}H]thymidine into the fibroblasts, and the concentration of virus in the cultures.

The inhibition of thymidine incorporation into the cells, caused by different HPE concentrations acting for one hour, was determined first of all. After twenty-four hours incorporation of [^{3}H]thymidine a dose–effect curve was obtained, and this is presented in figure 3.13. One can see that incorporation of thymidine into the DNA of the fibroblasts is inhibited to 50% by a concentration of 0·1 mM HPE, whereas half that concentration (50 μM) produces no significant alteration of thymidine incorporation. The controls were incubated in 0·9% sodium chloride solution. The half-maximal HPE concentration (0·1 mM) indicates a relatively high sensitivity of DNA synthesis in chick fibroblasts. This result fits well into the hypotheses so far developed on the connection between proliferative activity and sensitivity to HPE of DNA synthesis. With EATC, possessing an average doubling time of two days, half-maximal inhibition occurred with HPE concentrations between 60 μM and 0·1 mM. Chick fibroblasts proliferate distinctly more weakly with the doubling period being between three and four days (see figure 3.11).

Testing the viability of the cells with the Trypan Blue test showed that no alteration in the percentages of Trypan-positive cells (compared with

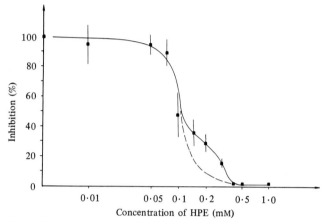

Figure 3.13. Inhibition of [^{3}H]thymidine incorporation into cultures of chick fibroblasts, by different HPE concentrations (according to Dostal et al.[202] and Kulnigg[203]). Controls, without HPE, represent 100%. The broken line shows the theoretically expected (calculated) curve.

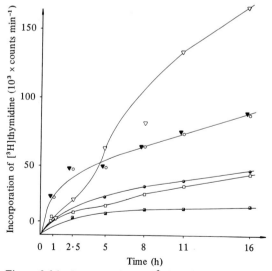

Figure 3.14. Incorporation of [^{3}H]thymidine into chick fibroblasts (according to Dostal et al.[202] and Kulnigg[203]). ▼, controls; ○, preincubated in 50 μM HPE; △, infected with vaccinia virus; □, virus infected, and incubated in 50 μM HPE; ●, preincubated in 0·1 mM HPE; ■, infected, and incubated in 0·1 mM HPE.

the controls) took place up to an HPE concentration of 0·15 mM. Morphological examination of the cells under the microscope also showed that up to concentrations of 0·15 mM HPE no subjectively assessed damage could be observed. Summarizing, then, it was observed that chick fibroblasts gave no indication of damage to viability after incubation in HPE at concentrations of ≤0·15 mM, even though DNA synthesis is already fifty percent inhibited at concentrations of 0·1 mM HPE. Figure 3.14 plots the incorporation of [^{3}H]thymidine into the cells, measured between zero time and sixteen hours.

If fibroblasts are infected with vaccinia virus a drastic alteration in the incorporation of thymidine results. Within the first five hours an approximately 30% decrease in the DNA synthesis occurs, which is then followed by a considerable stimulation, amounting to practically 100% of the control value after sixteen hours. Incubation of the infected cells in 50 μM HPE produces results represented by the points on the curve (figure 3.14) designated □, which correspond to a strong inhibition of thymidine incorporation into the cells. Relative to the control curve this amounts to a 72% inhibition after a sixteen hour incorporation period; after incubation in 0·1 mM HPE solution the inhibition is 86%. It still had to be established if this inhibition of thymidine incorporation might

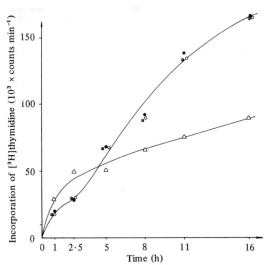

Figure 3.15. Incorporation of [^{3}H]thymidine into chick fibroblasts (according to Dostal et al.[65] and Kulnigg[66]). △, controls; ○, infected, virus preincubated in 0·9% sodium chloride; ◐, infected, virus preincubated in 50 μM HPE; ●, infected, virus preincubated in 0·1 mM HPE.

Plate 1. Effect of HPE on the solid Ehrlich tumour, peritumoral injection, three weeks after implantation. (a) untreated control, (b) 'delayed', (c) 'prevented' (according to Schauenstein et al.[194]).

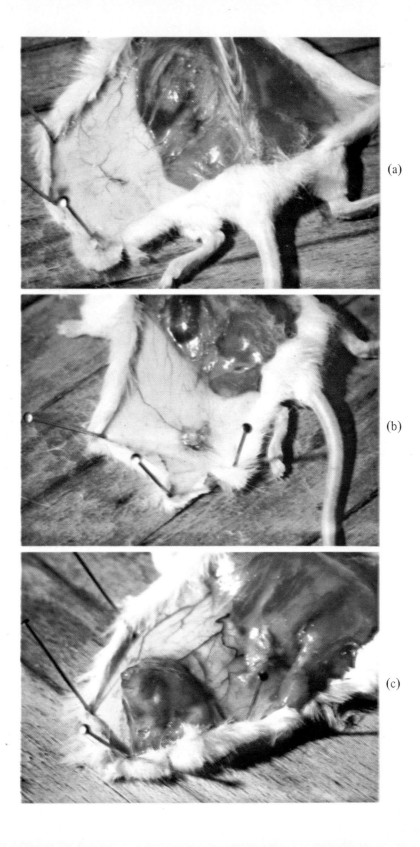

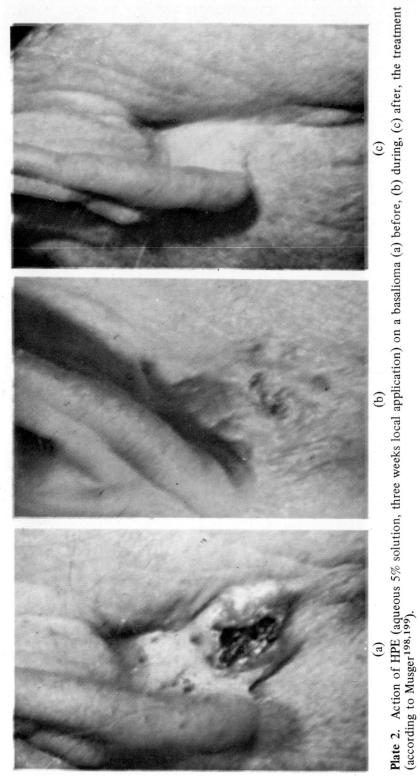

Plate 2. Action of HPE (aqueous 5% solution, three weeks local application) on a basalioma (a) before, (b) during, (c) after, the treatment (according to Musger[198,199]).

be caused by a deactivation of the virus particle by HPE. For this purpose three aliquots of a preparation of vaccinia virus were preincubated at 37°C for one hour: in 0·9% NaCl solution, in 0·1 mM HPE, and in 50 μM HPE. Fibroblast cultures were then infected with these aliquots of virus, and the incorporation of thymidine was measured. Figure 3.15 shows conclusively that the inhibitory effects are certainly not caused by a deactivation of the virus particle by HPE. The inhibitory effects, in particular at 50 μM HPE, must therefore be due to HPE selectively inhibiting only the stimulated incorporation of thymidine into the virus-infected cell. It may be concluded that we are dealing with a selective inhibition of viral DNA synthesis. Convincing evidence of this is given on determining the virus concentrations in the virus infected cultures when treated with HPE. Dostal et al.[202] found a decrease in the number of virus particles of 76%, on incubation in 50 μM HPE solution, and of 98% in 0·1 mM HPE. From the good correspondence between the inhibition of thymidine incorporation effected by HPE, and the reduction of virus concentrations in the infected fibroblasts, we may conclude that HPE blocks viral DNA synthesis. Since this occurs even at HPE concentrations (50 μM) which do not measurably affect DNA synthesis in the cell, we may be justified in speaking of a selective inhibition of viral DNA synthesis in the infected host cell.

The reason for this selective inhibition could be that viral DNA synthesis actually exhibits an activity many times higher than that peculiar to the cell itself (see figure 3.14). This is in agreement with previous concepts about the connection between synthetic activity and sensitivity towards HPE. The fact that two separate DNA syntheses take place simultaneously in one cell, and that inhibitions of them show these differences, demonstrates very clearly the selective activity of HPE. Basically this is true for all processes which are controlled by functional sulphydryl groups; the sensitivity towards HPE of such processes will be determined by the number of the controlling sulphydryl groups, or the chemical reactivity. The higher and specific performances of the cell, as for example muscle contraction, surely belong to such processes. It is known that the proteins of contractile muscles contain sulphydryl groups which are functionally important for the contraction process. In particular they are significant for the association of actin and myosin to form actomyosin, for the transformation of the globular form of actin into the fibrillar form, and for the ATPase activity of myosin; this latter process possesses the greatest sensitivity towards sulphydryl reagents. The influence that HPE exerts on smooth muscle (uterus), and striated muscle (diaphragm), of the rat, has been investigated in collaboration with Burkl et al.[204]. With rising HPE concentrations, the contractions of the uterus decrease in extent and finally stop completely, whereas the tonus is not affected (figures 3.16 and 3.17). With as little as 0·1 mM HPE, spontaneous activity ceases, but only after 40–60 minutes action; the activity can be reestablished by washing out the inhibitory agent.

With the diaphragm, contractions take place at still lower HPE concentrations (60 μM). This effect is not remediable by washing.

The effects referred to are surely based on the reaction of HPE with the above-mentioned functional sulphydryl groups, whose reactivity must be especially high, in any case much higher than that of the sulphydryl groups of functional importance for the respiration and glycolysis of these muscle preparations. Both organs show practically no inhibition of

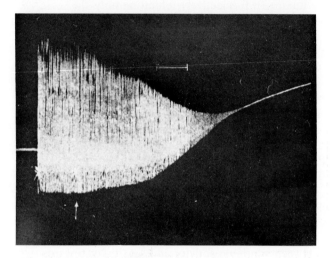

↑ addition of HPE
⊢⊣ 5 minutes

Figure 3.16. Action of HPE (2 mM) on electrically stimulated rat diaphragm (according to experiments of Burkl et al.[204]).

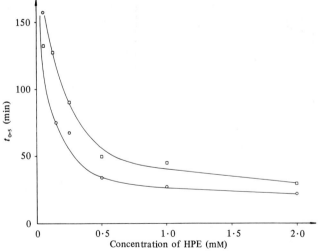

Figure 3.17. Effect of concentration of HPE on the time interval ($t_{0.5}$) between addition of HPE and half-maximal contractions of rat diaphragm (according to Burkl et al.[204]). □, resting muscle; ○, electrically stimulated muscle.

respiration or of glycolysis with HPE concentrations of 50 μM. One can therefore recognize the undoubted existence of an activity of HPE which is highly selective for different processes. In the final analysis this selectivity is dependent on the reactivity of the particular sulphydryl groups which control the appropriate process. HPE will always inhibit most effectively that process which is controlled by the sulphydryl groups having the highest reactivity.

The reactivity of the corresponding sulphydryl groups can be determined by the following factors:
(1) primary, secondary, and higher structures of proteins, into which the groups are built;
(2) accessibility within the cell (dependent on compartmentalization, permeability of membranes, etc.);
(3) protective substances (NPSH, in particular GSH).

One can thereby understand the selectivity of action which HPE exhibits if it acts on several simultaneous and parallel processes controlled by sulphydryl groups in one particular cell type. The above concepts also explain the selectivity of HPE towards one particular sulphydryl-group controlled process in different cell types. It also follows quite unambiguously that, for HPE, and for the HE in general, no *absolute* selectivity, that is, no actual specificity, is possible. The assessment of selective activity presupposes an exact definition of a quite precise cellular property which is selectively influenced by HPE, and also of those cell types which are available for comparison. Thus it would not be correct to say, for example, that HPE acts selectively on tumour cells. The proper statements would be that HPE damages the DNA synthesis of sarcoma-37 or Ehrlich tumour cells many times more strongly than that of cells from healthy liver, kidney, or spleen; or that HPE blocks three times as many sulphydryl groups of the soluble proteins in rapidly growing tumours as in those of slowly growing tumours. Only in this way can misinterpretations be avoided.

It is of considerable interest that another member of the group of 4-hydroxyenals is formed from another biological substrate, namely aqueous solutions of 2-deoxyribose.

Esterbauer *et al.*[204a] reported on the isolation and identification of 5,4-dihydroxy-pent-3,2-trans-en-1-al from unbuffered aqueous solutions of 2-deoxyribose (1%) after heating for varying numbers of hours (2 to 550) at temperatures up to 120°C. The substance is obtained with a maximal yield of 12·8% after heating at 120°C for 4 hours.

Plainly, the only difference from HPE is the presence of one additional hydroxy group at carbon atom 5. Because of this it was assumed that both compounds might show similar biochemical properties and cytological effects.

In fact Esterbauer *et al.*[204a] report that the substance is highly toxic and has a LD_{50} for mice amounting to 1·2–1·5 mg/mouse. Information on further biological effects is still lacking. However, one is tempted to consider the possibility that the dihydroxypentenal may also be present in native cells, and where it may function as a biogenic control for cell division, as Szent-Györgyi and Együd have postulated in a similar way for α-keto-aldehydes (see section 5.1.5.2).

3.3 Metabolic fate of α,β-unsaturated aldehydes

α,β-Unsaturated carbonyl compounds are in principle attacked metabolically both at the carbon–carbon double bond, which is activated by conjugation, and at the carbonyl group. If a further functional group is present, as for instance the hydroxyl group at C-4 of the hydroxyenals (sections 3.2.4 and 3.2.5), a further point of attack is created for metabolic reactions. Reactions at the carbon–carbon double bond may involve addition reactions or oxidation–reduction processes; for the carbonyl group the latter processes tend to predominate.

For reactions *in vivo* the additions of sulphydryl groups of glutathione to carbon–carbon double bonds conjugated to a carbonyl group have to be considered. According to the extensive investigations of Boyland and his research group, rat liver contains specific glutathione-S-transferases, which catalyse the transfer of the glutathione-S-residue to different substrates[205-209]. In further investigations Boyland and Chasseaud[210] described a glutathione-S-transferase which transfers the glutathione-S-group to α,β-unsaturated esters such as diethyl maleate. The glutathione-S-group may be transferred in the liver to α,β-unsaturated aldehydes, ketones, lactones, carboxylic acids, etc. It was established by chromatographic procedures that the corresponding substrate forms conjugates with glutathione, and that loss in glutathione, determined experimentally, is not caused by oxidation.

A comparison of separate members of the above-mentioned groups of α,β-unsaturated carbonyl compounds leads to interesting conclusions about the influence of chemical constitution of the substrate on its reactivity towards glutathione and the activity of the glutathione-S-transferase. Thus it was shown that substitution of a methyl group for hydrogen at the β-carbon atom strongly decreases the activity of the enzyme. A similar effect was observed on introducing an oxygen atom between the carbonyl carbon and α-carbon atoms, as in vinyl acetate. On the other hand, substitution of ester groups at α- or β-carbon atoms causes activation. The strongest activation occurs with nitroalkenes.

It is evident that these results are of fundamental importance for the evaluation of biochemical and biological effects *in vivo* of α,β-unsaturated carbonyl compounds, for example, in respect of toxicity, carcinogenesis, and carcinostatic effects, and for the problems of resorption, transport, detoxification, and elimination, etc.

The enzyme catalysing the coupling of glutathione with diethyl maleate was found to have the highest activity in rat liver, and lower activities in the lung (approximately 30%), in kidney (approximately 20%), heart, spleen, and blood (each approximately 10%). The activities of the liver enzyme are very different for different animal species, and are moreover sex-dependent. Thus mice (CBA/H) exhibit the highest activity (female 26, male 18·4 units) among the species examined, guinea pigs about half of these activities, rats (Chester Beattie strain, female 9–11, male 8·7 units), pig and rabbit approximately 5 units. In man (1–2·6 units) one finds the lowest activities in the foetus (0·6–1·7 units).

In a further investigation Boyland and Chasseaud[211] established that glutathione-S-alkene transferases represent a particular group of enzymes each with well marked substrate specificity and differing in their activities in the different organs. They can be readily distinguished from glutathione-S-alkyl, -aryl, and -epoxy transferases.

Whereas other detoxification mechanisms, such as glucuronide formation or sulphate conjugation, require primary ATP-dependent formation of intermediates with a high group-transfer potential, this does not appear to be the case with the enzymatic coupling to glutathione.

Some substrates react also nonenzymatically and very rapidly with sulphydryl groups. The biological significance of GSH-alkene transferases then lies in the protection they afford to sulphydryl groups of important cellular enzymes from attack by α,β-unsaturated carbonyl compounds. This protective function is probably also the essential reason why 4-hydroxy-2,3-alkenals inhibit certain metabolic and synthetic steps controlled by sulphydryl enzymes, much more weakly in normal liver and kidney cells than they do, for instance, in tumour cells (cf section 3.2.5).

After Boyland et al. had shown that α,β-unsaturated carbonyl compounds form addition products in the liver with GSH, Speir and Barnsley[212] considered the possibility that α,β-unsaturated acyl-S-CoA esters present in normal metabolism reacted with glutathione in an analogous fashion. They could then lead to the formation of intermediates whose hydrolysis would afford S-carboxyalkylcysteine derivatives, as in reactions 3.1:

$$\text{GSH} + \text{CH}_3\text{CH}=\text{CHCSR} \rightarrow \text{CH}_3\text{CHCH}_2\text{CSR} \rightarrow \text{CH}_3\text{CHCH}_2\text{COH} \quad 3.1$$

(with substituents: $\|O$ on CSR; $|SG$ and $\|O$ on middle product; $|S-CH_2-H_2NCHCOOH$ and $\|O$ on final product)

At this point Speir and Barnsley[212] succeeded in partially purifying an enzyme from the soluble fraction of rat liver. It catalyses the reaction of GSH with 2,3-unsaturated acylthiol esters. The enzyme is detectable only in the residue after centrifugation for one hour at 100 000 g of the liver homogenate. The nuclear, mitochondrial, or microsomal fractions showed no significant activity. The test system contains α,β-unsaturated acylthiol esters, GSH, and enzyme extract. Decreases of GSH and of ester, occurring in active extracts, are stoichiometrically related to each other. Both decreases agree well with the quantities of reaction products formed. Believed to be S-(2-carboxy-l-alkyl)-GSH, they were determined by paper chromatography and electrophoretically as their hydrolysis products, that is, as S-(2-carboxy-l-alkyl)cysteines. To be sure, as the authors themselves have remarked, the chemical identification of the reaction products is not absolutely established by these means. The pH optimum of the purified enzyme extract was estimated to be pH 8–9. K_m and V_{max} values differ considerably for the various α,β-unsaturated acyl-S-esters used. The enzyme is not identical with the GSH-S-alkene transferases found by Boyland et al., since no catalytic effect could be observed on the interaction between GSH and α,β-unsaturated carboxylic acids or their alkyl esters. The enzyme could be responsible for the formation of certain S-carboxyalkylcysteines found in animal urine, on the basis of the activities and specificities determined. Whether α,β-unsaturated acyl-S-CoA esters have such a metabolic fate in vivo remains a problem. On the one hand they are used up in normal degradation steps, but on the other hand they are compartmentalized in the mitochondria although the enzyme itself occurs in the cytoplasm. Nothing is known about whether the above CoA derivative reaches the cytoplasm from inside the mitochondria, and the manner in which this might be possible. In anomalous situations the enzyme could indeed play a role.

As already noted at the beginning of this section, α,β-unsaturated carbonyl compounds are also involved in oxidation–reduction processes of intermediary metabolism.

According to Williams[213], enzymatic reductions of the carbonyl group predominate. Other reductions have also been described in which both the carbon–carbon double bond and the carbonyl group of a substrate are involved[214], as well as enzymatic metabolic reactions in which it is only either the carbon–carbon double bond or the carbonyl function that is reduced[215]. The processes mentioned are catalysed by oxidoreductases requiring NADP or NAD as coenzyme, and are mostly localized in the liver.

Hardinge et al.[216] describe an enzyme which is present in the erythrocytes of dogs, and which is responsible for the rapid disappearance of furfurylideneacetone in heparinized dog blood; the enzyme is also found in liver. In a subsequent paper[217] the purification and characteristics of this enzyme from animal and human blood, liver, and kidney, are described. One is obviously dealing with a NADP-dependent oxidoreductase, which catalyses with high specificity the reduction of α,β-unsaturated ketones (such as benzylideneacetone, furfurylideneacetone, etc.) at an optimum pH of 5–7. Sulphydryl inhibitors, such as N-ethylmaleimide and p-chloromercuribenzoate, inhibit the enzyme, thus demonstrating that it is a sulphydryl group functional enzyme. Substrates possessing only a carbon–carbon double bond or only a carbonyl group, are not reduced. The measurements elegantly furnish two pieces of information on the progress and the mechanism of the enzyme-catalysed reaction. The intensity of absorption at 340 nm due to NADPH decreases, and also the absorption of the α,β-unsaturated substrate itself decreases. Determinations of the light absorptions are readily made in the quartz middle UV region, for example, furfurylideneacetone at 317 nm, or benzylideneacetone at 287 nm. Since the decreases of the two bands are stoichiometrically interrelated 1:1, it has been proved that the substrate has really undergone reduction, and has not, say, reacted with a sulphydryl group at the double bond.

Another enzyme has therefore been found, additional to the GSH-S-alkene transferases, which metabolizes α,β-unsaturated carbonyl compounds. The stoichiometric relation of 1:1 between the decreases in extinction at 340 nm and 287 nm, respectively, shows further that only one of the two reactive groups in benzylideneacetone (carbon–carbon or carbon–oxygen double bond) has been reduced, but allows no identification of the group reduced, since either of the two types of reductions would destroy the chromophoric system. Since NADPH participates it appears probable that it is the carbonyl group which is attacked, and is reduced to the corresponding alcohol.

It is remarkable that α,β-unsaturated carbonyl compounds of known high sulphydryl group reactivity do not block the functional (and highly reactive) sulphydryl groups of the oxidoreductase. Instead they are rapidly reduced.

The metabolic fate of the α,β-unsaturated carbonyl compounds which we have discussed so far is similar to that of the 4-hydroxyenals. Additional results obtained by Esterbauer et al.[219] supplement the investigations referred to earlier, insofar as they prove for the first time the chemical constitution of several metabolic products obtained in vivo.

(a) 4,5-^{14}C-labelled HPE diethyl acetal was given systemically to rats by intraperitoneal injection in a dose of 120 mg/kg body weight. Table 3.33 lists the activities found in the urine, blood, and various organs after fifteen minutes, and after twenty-four hours. Without doubt the main

Table 3.33. Radioactivity in organs of Wistar rats (250 g) after intraperitoneal injection of ^{14}C-labelled HPE diethylacetal in percentage of dose given (according to Esterbauer et al.[219]). Values refer to 1 g wet weight.

Organ	15 min	24 h	Organ	15 min	24 h
Blood	1·3	0·13	Heart	0·6	–
Liver	1·1	0·15	Striated muscle	~0·5	~0·13
Kidney	1·1	0·16	Total urine	–	62·0
Lung	0·8	0·13	Total faeces	–	3·4
Spleen	0·5	–			

part of the activity is eliminated in the urine. Separation of the dry residue of the twenty-four hour urine on DEAE-Sephadex gave five active components having different R_F values. No unchanged acetal remained (cf figure 3.18). Esterbauer et al. isolated the main components 2a and 2b, which gave only a single peak on the ion-exchange column but two spots by thin-layer chromatography. Separation into the two components 2a and 2b was achieved by chromatography on kiesel-gel, quantities of 12–13 mg of each being isolated. Other methods of investigation comprised titrations with sodium hydroxide, reaction with DNPH, polarimetry, and NMR, IR, and UV spectrometry. The positive conclusion based on these measurements is that components 2a and 2b are the stereoisomeric mercapturic acids, N-acetyl-S-(D-1-diethoxymethyl-3-oxobutyl)-L-cysteine, and N-acetyl-S-(L-1-diethoxymethyl-3-oxobutyl)-L-cysteine respectively. It follows that the diethyl acetal is metabolized in vivo, in agreement with the results of Boyland et al., mainly according to scheme 3.10. The main site for the metabolic processes represented should be liver and kidney, on the basis of the results of Boyland et al. quoted above, and the data given in table 3.33. Esterbauer et al. suppose that the first metabolic step consists of dehydrogenation of the alcoholic grouping sited at C-4. It is probable that the NADP-dependent oxidoreductase of the liver, as found by Frazer and coworkers[216-218], is responsible for this step. The α,β-unsaturated ketone thus formed represents a substrate for the GSH-S-alkene transferases according to Boyland. The fact that equal quantities of the diastereoisomeric metabolites 2a and 2b were found in the urine indicates that the appropriate alkene transferase is not stereospecific. The last step, N-acetylation of the α-amino group of the cysteine, is a reaction known for a long time to be responsible for the elimination of numerous substances foreign to the body.

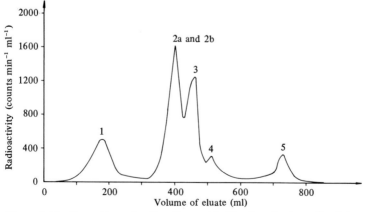

Figure 3.18. Separation of radioactive metabolites in rat urine, after intraperitoneal injection of ^{14}C-labelled HPE acetal, on DEAE-Sephadex A25 (according to Esterbauer et al.[219]).

Finally, Esterbauer[219] was able to confirm unambiguously the chemical constitution of metabolites 2a and 2b by synthesis (scheme 3.11).

Separation of the isomers, like that of metabolites 2a and 2b, was achieved on a column of kiesel-gel with chloroform–formic acid. The substance which, on thin-layer chromatographic separation (kiesel-gel; chloroform–acetic acid 5:1), had the higher R_F value was denoted 2a; the substance with the lower R_F value was denoted 2b.

Comparisons between the urine metabolite 2a and the synthetic substance 2a, or of metabolite 2b with the synthetic substance 2b, showed in each case identical IR and NMR spectra, pK values, R_F and $α_D$ values, and UV absorption maxima of the 2,4-dinitrophenylhydrazones in methanol and potassium hydroxide.

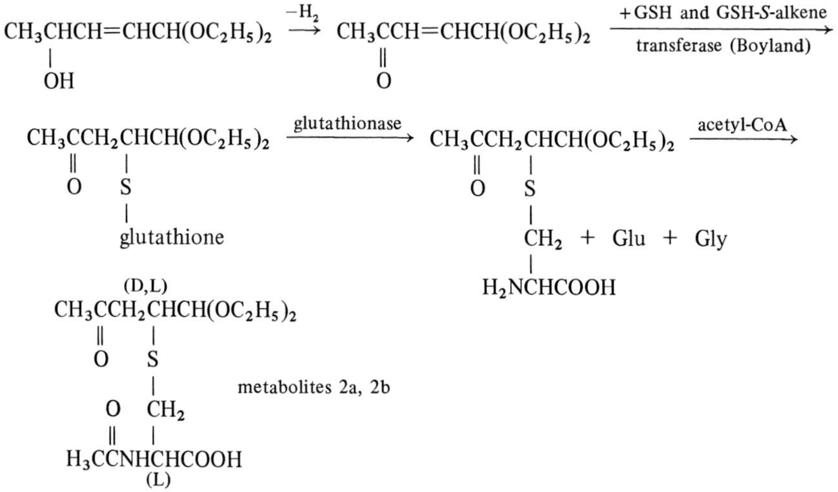

$$CH_3CHCH=CHCH(OC_2H_5)_2 \xrightarrow{-H_2} CH_3CCH=CHCH(OC_2H_5)_2 \xrightarrow[\text{transferase (Boyland)}]{+GSH \text{ and } GSH\text{-}S\text{-alkene}}$$
(with OH on first carbon; O= on second structure)

$$CH_3CCH_2CHCH(OC_2H_5)_2 \xrightarrow{\text{glutathionase}} CH_3CCH_2CHCH(OC_2H_5)_2 \xrightarrow{\text{acetyl-CoA}}$$
(with =O and S–glutathione; then =O and S–CH$_2$ + Glu + Gly, H$_2$NCHCOOH)

(D,L)
$$CH_3CCH_2CHCH(OC_2H_5)_2$$
(with =O, S–CH$_2$, O=C(CH$_3$)NHCHCOOH (L))

metabolites 2a, 2b

Scheme 3.10

$$CH_3CHC\equiv CCH(OC_2H_5)_2 \xrightarrow{LiAlH_4} CH_3CHCH=CHCH(OC_2H_5)_2 \xrightarrow[\text{acetone}]{Al,\ t\text{-butoxide}}$$
(OH on first; OH on second)

$$CH_3CCH=CHCH(OC_2H_5)_2 + HSCH_2CHCOOH \rightarrow CH_3CCH_2CHCH(OC_2H_5)_2$$
(with =O; NHCOCH$_3$; and product with =O, S–CH$_2$, CH$_3$C(=O)NHCHCOOH)

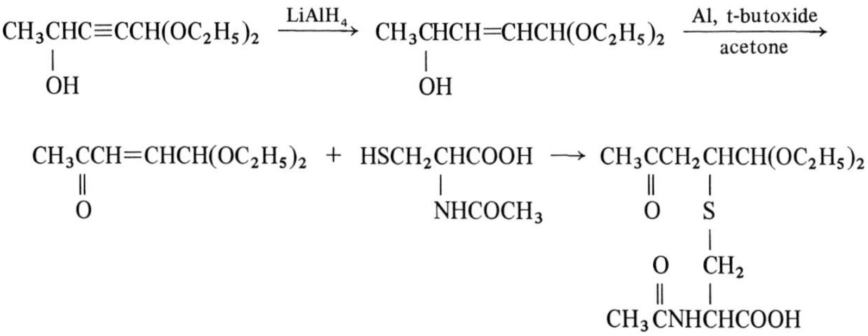

1:1 mixture of D,L- and L,L-form, 2a, 2b

Scheme 3.11

The first exploratory investigations of metabolite 1 showed that two substances were eluted simultaneously in elution fraction 170, which were separated by thin-layer chromatography. One of these components had the same R_F value as synthetic 4-oxopentenal diethyl acetal and is thus probably identical with the first intermediate product given in scheme 3.10. (b) Initial results obtained by Esterbauer (unpublished) also exist for the metabolic fate *in vivo* of free 4-hydroxypentenal. Of 150000 counts min^{-1} injected intraperitoneally (16·4 mg of 4,5-^{14}C-labelled HPE) about 60% were found in the urine after twenty-four hours, corresponding to the results with HPE diethyl acetal. Separation of the urine components on DEAE-Sephadex A25 proceeded as for those from the acetal, and the elution diagram is reproduced in figure 3.19. At least four metabolites were recognized, whose elution fractions corresponded to those of the metabolites of the diethyl acetal. Table 3.34 gives the proportions of these metabolites, which for metabolites 2, 3 and 5 show distinct

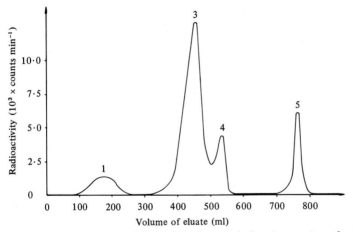

Figure 3.19. Separation of the radioactive metabolites in rat urine, after intraperitoneal injection of ^{14}C-labelled HPE, on DEAE-Sephadex A25 (according to H. Esterbauer, unpublished experiments).

Table 3.34. Metabolites found in the 24 h urine of rats after intraperitoneal injection of HPE and HPE diethyl acetal respectively (according to H. Esterbauer, unpublished).

Metabolite	HPE acetal		Free HPE	
	%	V_e	%	V_e
1	11	170	8	170
2	44	400	0	400
3	26	460	64	460
4	9	515	8	530
5	7	730	20	760

differences between HPE and HPE diethyl acetal. Metabolite 2, found only in the case of the acetal, consisted of the two substances discussed above (cf page 92).

Because the fractions correspond fairly well it may be argued that the urine metabolites 3, 4, and 5 are the same as in the case of the HPE acetal. (c) The investigations of Binder[220] provide information on the further metabolic fate of 4-hydroxypentenal when acting on normal rat liver *in vitro*[1]. If freshly excised normal rat liver (in pieces of approximately 1 g wet weight) is incubated in isotonic HPE solution at room temperature, the HPE in the medium is depleted with increasing duration of the incubation (see figure 3.20). This result appears reasonable in view of the presence in the liver of GSH-S-alkene transferases, as described above, as well as of NADP-dependent oxidoreductases specific for α,β-unsaturated carbonyl compounds. Thin-layer chromatography of the incubation medium, with diethyl ether as developing solvent, gave, besides a pronounced spot (R_F 0·8) corresponding to residual HPE, another spot (R_F 0·34) which increased with increasing incubation time. This spot also occurs on incubation of normal rat kidney in HPE with, however, much lower intensity. On incubation of heart, spleen, and lung of healthy rats it is barely detectable. The spot is completely missing when the incubation medium is free from HPE. It may be concluded that the compound yielding the spot of R_F 0·34 is also a product of an alcohol–aldehyde oxidoreductase, that is, 1,4-dihydroxy-2-pentene. This compound was therefore also tested by thin-layer chromatography and was found to give identical R_F values—the first concrete hint towards the subsequent isolation and identification of the material. Success was actually achieved

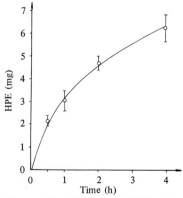

Figure 3.20. Uptake of HPE (mg) by 1 g of fresh rat liver on incubation in 8 ml of isotonic (50 mM) HPE solution (equivalent to 40 mg of HPE) at room temperature (according to Binder[220]).

[1] Esterbauer will report on these investigations in detail, in a comprehensive presentation of the metabolic fate of HPE *in vivo* and *in vitro*.

by using the 4′-nitroazobenzenecarboxylic acid ester (NABA-ester). Fresh liver (25 g) was incubated for four hours at 20°C in 153 ml of isotonic (50 mM) HPE solution; the mixture was filtered, and the filtrate reduced to one-tenth of its volume; the components were then separated on Sephadex G25. The desired substance was isolated by multiple distribution, using thin-layer chromatographic control, 21 mg of pure compound being finally obtained. Its IR absorption spectrum corresponded completely with that of synthetic 1,4-dihydroxypentene. The substance was then converted into its NABA-ester, being identified UV spectrometrically as such after purification by recrystallization. The IR spectrum was also identical with that of the NABA-ester of synthetic 1,4-dihydroxypentene, as was the NMR spectrum. The metabolic product was therefore concluded to be 1,4-dihydroxy-2-pentene.

Experiments were subsequently undertaken on the supposed biogenesis of 1,4-dihydroxypentene from 4-hydroxypentenal in normal rat liver slices *in vitro*. These experiments showed clearly that an enzyme-catalysed reaction was involved. The responsible enzyme could be provisionally identified as a sulphydryl group functional NAD-dependent alcohol oxidoreductase, which is known to reduce a large number of aldehydes[221], and evidently acetaldehyde much more readily than hydroxypentenal. This can be clearly seen from the chromatogram (d) depicted in figure 3.21.

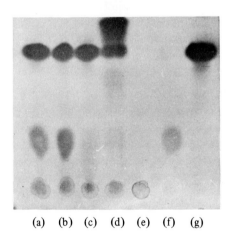

(a) (b) (c) (d) (e) (f) (g)

Figure 3.21. Thin-layer chromatographic examination of the following incubation media:
(a) 2 ml of HPE (10 mg/ml), 2 ml of enzyme extract from rat liver, and 10 mg of $NADH_2$ at 20°C for 2 h;
(b) 2 ml of HPE [as (a)], 2 ml of enzyme extract [as (a)], and 10 mg of $NADH_2$ at 37°C for 2 h;
(c) 2 ml of HPE [as (a)], and 2 ml of enzyme extract [as (a)] at 37°C for 2 h;
(d) 2 ml HPE [as (a)], 2 ml of enzyme extract [as (a)], acetaldehyde (0·25 mM), and 10 mg of $NADH_2$ at 37°C for 2 h;
(e) 2 ml of enzyme extract [as (a)], and 10 mg of $NADH_2$ at 37°C for 2 h;
(f) 1,4-dihydroxypentene; (g) HPE.

Thus the experiments demonstrated that HPE is reduced in the liver to the corresponding alcohol by an NAD-dependent alcohol oxidoreductase. The question which is now posed is whether this metabolic process also occurs *in vivo* and whether 1,4-dihydroxypentene may also be eliminated in the urine. As yet this question cannot be answered. The metabolites 3, 4, and 5 (cf pages 94, 95) occurring in the twenty-four hour urine after intraperitoneal injection of HPE are certainly not 1,4-dihydroxypentene, since they all carry negative charges. It is quite possible, however, that the substance is present in fraction 170.

References
1. Young, W. G., 1932, *J. Am. Chem. Soc.*, **54**, 2500.
2. Blacet, F. E., Young, W. G., 1937, *J. Am. Chem. Soc.*, **59**, 612.
3. Gredy, B., Piaux, L., 1934, *C. R. Acad. Sci.*, **198**, 1236.
4. Thomas, D. A., Warburton, W. K., 1965, *J. Chem. Soc.*, 2988.
5. Buswell, A. M., Dunlop, E. C., Rodebush, W. H., Swartz, J. B., 1940, *J. Am. Chem. Soc.*, **62**, 326.
6. Bunewag, A., 1941, *J. Chem. Soc.*, 1939.
7. McKinney, G., Temmer, O., 1948, *J. Am. Chem. Soc.*, **70**, 3887.
8. Blout, O. R., Fields, M., 1948, *J. Am. Chem. Soc.*, **70**, 191.
9. Moskvin, A. F., Yablonski, O. P., Bonder, L. F., 1966, *Teor. Eksp. Khim.*, **2**, 635 [*Chem. Abstr.*, **66**, 120376 (1967)].
10. Esterbauer, H., Weger, W., 1967, *Monatsh. Chem.*, **98**, 1884, 1994.
11. Le Henaff, P., 1967, *C. R. Acad. Sci.*, **265**, 175.
12. Polansky, O., 1957, *Monatsh. Chem.*, **88**, 91.
13. Brown, R., Jones, W. E., Pinder, A. R., 1951, *J. Chem. Soc.*, 3315.
14. Cuneen, J. I., Ford, J., 1956, *J. Chem. Soc.*, 3056.
15. Friedman, M., Wall, J. S., 1964, *J. Am. Chem. Soc.*, **86**, 3735.
16. Friedman, M., Cavins, J. F., Wall, J. S., 1965, *J. Am. Chem. Soc.*, **87**, 3672.
17. Cavins, J. F., Friedman, M., 1968, *J. Biol. Chem.*, **243**, 3357.
18. Narimanbekov, O. A., Ismailov, P., Nektiev, S. D., 1967, *Dokl. Akad. Nauk SSSR*, **23**, 15 [*Chem. Abstr.*, **68**, 67178 (1968)].
19. Esterbauer, H., 1972, *Monatsh. Chem.*, **101**, 782.
20. Scholz, N., 1972, Dissertation, University of Graz.
20a. Friedman, M., 1973, *The Chemistry and Biochemistry of the Sulfhydryl Group in Amino Acids, Peptides and Proteins* (Pergamon Press, Oxford).
21. Fernandez, J. E., Solomons, T. W., 1962, *Chem. Rev.*, **62**, 485.
22. Friedman, M., 1967, *J. Am. Chem. Soc.*, **89**, 4709.
23. Cavins, J. F., Friedman, M., 1967, *Biochemistry*, **6**, 3766.
24. Bose, S. M., Thomas, K., 1957, *J. Am. Leather Chem. Assoc.*, **52**, 200.
25. Milch, R. A., Frisko, L. J., Szymkoviak, E. A., 1965, *Biorheology*, **3**, 9.
26. Travska, Z., Sitaj, S., Gremla, M., Malinsky, J., 1966, *Biochim. Biophys. Acta*, **126**, 373.
27. Bowes, J. H., Cater, C. W., 1968, *Biochim. Biophys. Acta*, **168**, 341.
28. Milch, R. A., 1963, *Gerontologia*, **7**, 129.
29. Milch, R. A., 1965, *South. Med. J.*, **58**, 153.
30. Deyl, Z., Rosmus, J., 1963, *Kozarstvi*, **13**, 139.
31. Milch, R. A., Clifford, R. E., Murray, R. A., 1966, *Nature (London)*, **210**, 1042.
32. Milch, R. A., 1965, *J. Atheroscler. Res.*, **5**, 215.
33. Mayor, H. D., Jordan, L. E., 1963, *J. Cell Biol.*, **18**, 207.
34. Feder, N., Wolf, M. K., 1965, *J. Cell Biol.*, **27**, 327.

35 Feustel, E. M., Geyer, G., 1966, *Acta Histochem.*, **25**, 219.
36 Halbhuber, K. J., 1966, *Acta Histochem.*, **24**, 385.
37 Alarcon, R. A., 1970, *Arch. Biochem. Biophys.*, **137**, 365.
38 Stack, V. T., 1957, *Ind. Eng. Chem.*, **49**, 913.
39 Zollner, H., Esterbauer, H., 1973, unpublished experiments.
40 Schauenstein, E., Esterbauer, H., Zollner, H., Kollaritsch, K., 1973, unpublished experiments.
41 Kollaritsch, K., 1973, Dissertation, University of Graz.
42 Williams, R. T., 1959, *Detoxication Mechanisms*, 2nd edn. (Chapman and Hall, London), pp.60-97.
43 Sim, V. M., Pattle, R. E., 1957, *J. Am. Med. Assoc.*, **165**, 1908.
44 Plotnikova, M. M., 1960, *Predel'no Dopustimye Konts. Atmos. Zagryaz.*, **4**, 75 [*Chem. Abstr.*, **55**, 12721 (1961)].
45 Soriano, M., 1958, *Sem. Hop.*, **34**, 3213 [*Chem. Abstr.*, **54**, 1713 (1960)].
46 Schaborth, J. C., 1967, *J. S. Afr. Chem. Inst.*, **20**, 103.
47 Murphy, S. D., Davis, H. V., Zaratzian, V. L., 1964, *Toxicol. Appl. Pharmacol.*, **6**, 520.
48 Murphy, S. D., 1965, *Toxicol. Appl. Pharmacol.*, **7**, 833.
49 Gusev, M. I., Dronov, I. S., Svechnikova, A. I., Golovina, A. I., Grebenskova, M. D., 1967, *Biol. Deistvie Gig. Znach. Atmos. Zagryaz.*, **10**, 122 [*Chem. Abstr.*, **69**, 21748 (1968)].
50 Grebenskova, M. D., 1965, *Dokl. ko 2-oi Ob'edin. Nauchn. Konf. Med. i. Nauchn.-Issled. Inst. Rostov on Don*, Pt 1, 424 [*Chem. Abstr.*, **65**, 1281 (1966)].
51 Murphy, S. D., Porter, S., 1966, *Biochem. Pharmacol.*, **15**, 1665.
52 Riddik, J. H., De Lee, S., Coffin, D. L., 1968, *Am. Chem. Soc., Div. Water, Air Waste Chem., Gen. Pap.*, **8**, 148 [*Chem. Abstr.*, **72**, 1775 (1970)].
53 Ryazanov, V. A., 1960, *Proc. Int. Clean Air Conf. London 1959*, 175 [*Chem. Abstr.*, **55**, 13726 (1961)].
54 Sinkuviene, D., 1970, *Gig. Sanit.*, **35**, 6 [*Chem. Abstr.*, **72**, 136068 (1970)].
55 Goeva, O. E., 1965, *Tr. Leningr. Sanit. Gig. Med. Inst.*, **81**, 63 [*Chem. Abstr.*, **66**, 118699 (1967)].
56 Murphy, S. D., Klingshirn, D. A., Ulrich, C. E., 1963, *J. Pharmacol. Exp. Ther.*, **141**, 71.
57 Portius, H. J., Repke, K., 1964, *Arzneimittel Forsch.*, **14**, 1073.
58 Izard, C., 1967, *C. R. Acad. Sci. Ser. D*, **265**, 1199.
59 Puiseux-Dao, S., Izard, C., 1968, *C. R. Acad. Sci. Ser. D*, **267**, 74.
60 Otto, N. E., Bartley, T. R., 1966, *U.S. Govt. Res. Develop. Rept.*, **41**, 93 [*Chem. Abstr.*, **67**, 10608 (1967)].
61 Rees, K. R., Tarlow, M. J., 1967, *Biochem. J.*, **104**, 757.
62 Philippin, K., 1969, *Praeventivmedizin*, **14**, 317.
63 Champeix, J., Pierre, C., 1967, *Poisoning by Acrolein* (Masson, Paris).
64 Skog, E., 1950, *Acta Pharmacol. Toxicol.*, **6**, 299.
65 Rinehart, W. E., 1967, *Am. Ind. Hyg. Assoc. J.*, **28**, 561.
66 Trovimov, L. V., 1962, *Gig. Tr. Prof. Zabol.*, **6**, 34 [*Chem. Abstr.*, **58**, 868 (1963)].
67 Boyland, E., 1940, *Biochem. J.*, **34**, 1196.
68 Motycka, K., Iacko, L., 1965, *Z. Krebsforsch.*, **66**, 491.
69 Motycka, K., Iacko, L., 1966, *Z. Krebsforsch.*, **68**, 195.
70 Boyland, E., Mawson, J., 1938, *Biochem. J.*, **32**, 1982.
71 Dittmar, C., 1940, *Z. Krebsforsch.*, **49**, 515.
72 Osato, S., 1950, *Tohoku J. Exp. Med.*, **52**, 181.
73 Osato, S., Oda, K., Hanno, H., 1953, *Gann*, **44**, 348.
74 Osato, S., Oda, K., Kitamura, S., 1954, *Bull. Assoc. Fr. Etude Cancer*, **41**, 466.

75 Morrow, J. W., 1960, *Br. J. Urol.*, **32**, 69.
76 Osato, S., Mori, H., Morita, M., 1961, *Tohoku J. Exp. Med.*, **75**, 223.
77 Osato, S., 1965, *Tohoku J. Exp. Med.*, **86**, 102.
78 Zolotovick, G., Nachev, K., Silyanovska, K., Stoichev, S., 1967, *C. R. Acad. Bulg. Sci.*, **20**, 1081, 1213.
79 Leach, E. H., Lloyd, J. P. F., 1956, *Proc. Nutr. Soc.*, **15**, 15.
80 Herzmann, E., 1966, *Acta Biol. Med. Ger.*, **17**, 550.
81 Alarcon, R. A., Meienhofer, J., 1971, *Nature (London)*, **233**, 250.
82 Arnold, H., Bourseaux, F., 1958, *Angew. Chem.*, **70**, 539.
83 Brock, N., Hohorst, H. J., 1963, *Arzneimittel Forsch.*, **13**, 1021.
84 Hirsch, J. G., Dubos, R. J., 1952, *J. Exp. Med.*, **95**, 191.
85 Hirsch, J. G., 1953, *J. Exp. Med.*, **97**, 345.
86 Kimes, R. W., Morris, D. R., 1971, *Biochim. Biophys. Acta*, **228**, 235.
87 Bachrach, U., Persky, S., 1964, *J. Gen. Microbiol.*, **37**, 195.
88 Alarcon, R. A., 1964, *Arch. Biochem. Biophys.*, **106**, 240.
89 Kimes, B. W., Morris, D. R., 1971, *Biochim. Biophys. Acta*, **228**, 223.
90 Ingersoll, R. L., 1938, *Food Res.*, **3**, 389.
91 Cavallito, C. J., Bailey, J. H., 1950, *J. Am. Chem. Soc.*, **66**, 1944.
92 Maatschappij, N. V., 1967, *Ger. Patent* 1.247-743, Aug. 17 [*Chem. Abstr.*, **68**, 4651 (1968)].
93 Osnos, G. M., 1964, *Lab. Delo*, **4**, 226 [*Chem. Abstr.*, **61**, 8620 (1964)].
94 Oh, H. K., Sakai, T., Jones, M. B., Longhurst, W. M., 1967, *Appl. Microbiol.*, **15**, 777.
95 Geiger, W. B., Conn, J. E., 1945, *J. Am. Chem. Soc.*, **67**, 112.
96 McGowan, J. C., Brian, P. W., Hemming, H. G., 1948, *Ann. Appl. Biol.*, **35**, 25.
97 Bosquet, E. W., Kirby, J. E., Searle, N. E., 1943, *U.S. Patent* No. 2335384 [*Chem. Abstr.*, **38**, 2834 (1944)].
98 Schildknecht, H., Rauch, G., 1961, *Z. Naturforsch.*, **16b**, 422, 301.
99 Tokin, B. P., 1956, *Phytoncide* (VEB Verlag Volk und Gesundheit, Berlin).
100 Major, R. T., Marchini, P., Boulton, A. J., 1963, *J. Biol. Chem.*, **238**, 1813.
101 Manowitz, M., Walter, G., 1964, *J. Pharm. Sci.*, **53**, 220.
102 Gingras, B. A., Colin, G., Bayley, C. H., 1965, *J. Pharm. Sci.*, **54**, 1674.
103 Viallier, J., Ollagnier, C., Augagneur, J., 1959, *C. R. Soc. Biol.*, **153**, 109.
104 Lespagnol, A., Gernez-Rieux, C., Beerens, H., Taquet, A., 1959, *Therapie*, **14**, 313 [*Chem. Abstr.*, **54**, 3745 (1960)].
105 Isler, J., Straub, O., 1956, *Swiss Patent* 315548 [*Chem. Abstr.*, **52**, 450 (1958)].
106 Bachrach, U., Rosenkovitch, E., 1972, *Appl. Microbiol.*, **23**, 232.
107 Kremzner, T. L., Harter, D. H., 1970, *Biochem. Pharmacol.*, **19**, 2541.
108 Eilon, G., Bachrach, U., 1969, *Biochim. Biophys. Acta*, **179**, 464.
109 Rapoport, S. M., 1948, *Dokl. Akad. Nauk SSSR*, **61**, 713 [*Chem. Abstr.*, **43**, 1115 (1949)].
110 Tiffany, B. D., Wright, J. B., Moffet, R. B., Heinzelman, R. C., 1957, *J. Am. Chem. Soc.*, **79**, 1682.
111 Goodhue, L. D., 1956, *U.S. Patent* No. 2770921 [*Chem. Abstr.*, **51**, 3902 (1957)].
112 Nyman, V. L., 1954, *U.S. Patent* No. 2690627 [*Chem. Abstr.*, **49**, 2013 (1955)].
113 Munsch, N., Frayssinet, C., 1971, *Biochimie*, **53**, 243.
114 Munsch, N., de Recondo, A. M., Frayssinet, C., 1973, *FEBS Lett.*, **30**, 286.
115 Rossignol, J. M., Abadicdebat, J., Tillit, J., de Recondo, A. M., 1972, *Biochimie*, **54**, 319.
116 Jovin, T. M., Englund, P. T., Kornberg, A., 1969, *J. Biol. Chem.*, **244**, 3009.
117 Benedict, R. C., Stedman, R. L., 1969, *Tob. Sci.*, **13**, 166.

118 Bilimoria, M. H., Nisbet, M. A., 1971, *Proc. Soc. Exp. Biol. Med.*, **136**, 698.
119 Battigelli, M. C., 1963, *J. Occup. Med.*, **5**, 54.
120 Kisarow, V. M., 1963, *Zavod. Lab.*, **29**, 163.
121 Zorin, V. M., 1966, *Gig. Sanit*, **31**, 98 [*Chem. Abstr.*, **66**, 31833 (1967)].
122 Laurene, A. H., Lyerly, L. A., Young, G. W., 1964, *Tob. Sci.*, **8**, 150.
123 Brock, N., Hohorst, H. J., 1967, *Cancer*, **20**, 900.
124 Hill, D. L., Laster, W. R., Struck, R. F., 1972, *Cancer Res.*, **32** 658.
125 Alarcon, R. A., Heienhofer, J., Atherton, E., 1972, *Cancer Res.*, **32**, 2519.
126 Phillips, B. J., *Second Meeting of the European Association for Cancer Research, Heidelberg, 1973*, Abstr. p.97.
127 Connors, T. A., Cox, P. J., Farmer, P. B., Foster, A. B., Jarman, M., 1974, *Biochem. Pharmacol.*, **23**, 115.
128 Schauenstein, E., Gold, O., Pibus, B., 1956, *Monatsh. Chem.*, **87**, 144.
129 Esterbauer, H., Schauenstein, E., 1967, *Die Nahrung*, **11**, 607.
130 Schauenstein, E., 1967, *J. Lipid Res.*, **8** 417.
131 Schauenstein, E., Biheller, J., 1956, *Monatsh. Chem.*, **87**, 158.
132 Schauenstein, E., Biheller, J., 1957, *Monatsh. Chem.*, **88**, 132.
133 Schauenstein, E., Schatz, G., 1959, *Fette, Seifen, Anstrichm.*, **61**, 1068.
134 Schauenstein, E., Schatz, G., Benedikt, G., 1961, *Monatsh. Chem.*, **92**, 442.
135 Schauenstein, E., Esterbauer, H., 1963, *Monatsh. Chem.*, **94**, 164.
136 Esterbauer, H., Schauenstein, E., 1963, *Monatsh. Chem.*, **94**, 998.
137 Klöpffer, W., Esterbauer, H., Schatz, E., 1965, *Fette, Seifen, Anstrichm.*, **67**, 198.
138 Esterbauer, H., Schauenstein, E., 1966, *Fette, Seifen, Anstrichm.*, **68**, 7.
139 Hubmann, A., 1964, Dissertation, University of Graz.
140 Schauenstein, E., Esterbauer, H., 1968, *Fette, Seifen, Anstrichm.*, **70**, 4.
141 Schauenstein, E., Bayzer, H., Krings, H., 1958, *Monatsh. Chem.*, **89**, 455.
142 Schauenstein, E., Schatz, G., 1959, *Monatsh. Chem.*, **90**, 118.
143 Schauenstein, E., Schatz, G., 1959, *Fette, Seifen, Anstrichm.*, **61**, 1068.
144 Schauenstein, E., Schatz, G., Benedikt, G., 1961, *Monatsh. Chem.*, **92**, 442.
145 Schauenstein, E., Schatz, G., Taufer, M., 1961, *Z. Krebsforsch.*, **64**, 465.
146 Zangger, J., Ratzenhofer, M., Schauenstein, E., 1961, *Z. Krebsforsch.*, **64**, 473.
147 Schauenstein, E., Taufer, M., Schatz, G., 1962, *Monatsh. Chem.*, **93**, 544.
148 Schauenstein, E., Zangger, J., Ratzenhofer, M., 1964, *Z. Naturforsch.*, **19b**, 923.
149 Jaag, G., Taufer, M., Schauenstein, E., 1964, *Monatsh. Chem.*, **95**, 1671.
150 Schauenstein, E., Jaag, G., Taufer, M., 1965, *Monatsh. Chem.*, **96**, 1485.
151 Jaag, G., 1964, Dissertation, University of Graz.
152 Auböck, L., 1966, Dissertation, University of Graz.
153 Dorner, F., 1967, Dissertation, University of Graz.
154 Schauenstein, E., Dorner, F., Sonnenbichler, J., 1968, *Z. Naturforsch.* **23b**, 316.
155 Schauenstein, E., Esterbauer, H., Jaag, G., Taufer, M., 1964, *Monatsh. Chem.*, **95**, 180.
156 King, D. W., Paulson, S. R., Puckett, N. L., Krebs, A. T., 1959, *Am. J. Pathol.*, **35**, 1067.
157 Ratzenhofer, M., Zangger, J., 1964, *Beitr. Pathol. Anat. Allg. Pathol.*, **130**, 243.
158 Esterbauer, H., Weger, W., 1967, *Monatsh. Chem.*, **98**, 1884.
159 Esterbauer, H., Weger, W., 1967, *Monatsh. Chem.*, **98**, 1994.
160 Esterbauer, H., 1971, *Monatsh. Chem.*, **102**, 824.
161 Ardenne, M. von, 1967, *Theoretische und Experimentelle Grundlagen der Krebsmehrschritt-Therapie* (VEB Verlag Volk und Gesundheit, Berlin), pp.307, 308, 313.
162 Burk, D., Woods, M., *Fifth International Congress of Chemotherapy, Vienna, 1967*, Congr. Rep. AIV-1b/13.

162a Conroy, P. J., Nodes, J. T., Slater, T. F., White, G. W., 1975, *Europ. J. Cancer*, **11**, 231.
163 Schauenstein, E., Wünschmann, B., Esterbauer, H., 1968, *Z. Krebsforsch.*, **71**, 21.
164 Kollaritsch, K., 1973, Dissertation, University of Graz.
165 Bickis, I. J., Schauenstein, E., Taufer, M., 1968, *Monatsh. Chem.*, **100**, 1077.
166 Bickis, I. J., Henderson, D., Quastel, J., 1966, *Cancer*, **19**, 103.
167 Schauenstein, E., Taufer, M., Esterbauer, H., Kylianek, A., Seelich, T., 1971, *Monatsh. Chem.*, **102**, 517.
168 Velick, S., Furfine, C., 1963, *The Enzymes*, volume 7, Eds P. D. Boyer, H. Lardy, K. Myrbäck (Academic Press, New York), p.243.
169 Bickis, I. J., unpublished work.
170 Roitt, J., 1965, *Biochem. J.*, **63**, 300.
171 Schmidt, C. G., 1964, *Acta Unio Int. Contra Cancrum. (Löwen)*, **20**, 968.
172 Schauenstein, E., Verdino, H., Taufer, M., 1970, *Monatsh. Chem.*, **101**, 1189.
173 Schauenstein, E., Kapfer, E., 1972, *Monatsh. Chem.*, **103**, 1200.
174 Bickis, I. J., Quastel, J., 1965, *Nature (London)*, **205**, 44.
175 Glock, G. E., MacLean, P., 1953, *Biochem. J.*, **55**, 400
176 Uppin, B. I., Scholefield, P. G., 1964, *Can. J. Biochem.*, **43**, 209.
177 Seeber, S., Warnecke, P., Weser, U., 1969, *Z. Krebsforsch.*, **72**, 137.
178 Anderson, L., 1966, *Biochim. Biophys. Acta*, **127**, 115.
179 Rindler, R., Schauenstein, E., 1970, *Z. Naturforsch.*, **25b**, 739.
180 Barrnett, R. J., Seligman, A. M., 1952, *Science*, **116**, 323.
181 Esterbauer, H., 1972, *Acta Histochem.*, **42**, 351.
182 Esterbauer, H., 1973, *Acta Histochem.*, **47**, 94.
183 Esterbauer, H., Nöhammer, G., Schauenstein, E., Weber, P., 1973, *Acta Histochem.*, **47**, 106.
184 Scheuringer, J., 1970, Dissertation, University of Graz.
184a Nöhammer, G., Schauenstein, E., Weber, P., 1973, *J. Histochem. Cytochem.*, **12**, 1082.
185 Burkl, W., Kramer, I., Schauenstein, E., 1967, *Z. Naturforsch.*, **22b**, 763.
186 Schindler, R., 1971, Dissertation, University of Graz.
186a Carow, H., 1976, Dissertation, University of Graz.
186b Schauenstein E., Schindler, R., 1976, quoted from: *Molecular Base of Malignancy*, Eds E. Deutsch, K. Moser, H. Rainer, A. Stacher (Thieme, Stuttgart) p.55
186c Schauenstein, E., 1975, *Int. J. Sulfur Chem.*, **9** (3), ...
187 Schauenstein, E., Rindler, R., Schindler, R., Taufer, M., 1971, *Z. Naturforsch.*, **26b**, 788.
188 Heise, E., Görlich, M., 1964, *Exp. Cell Res.*, **33**, 289.
189 Heise, E., Görlich, M., 1965, *J. Nat. Cancer Inst.*, **35**, 413.
190 Schauenstein, E., Höfler-Bergthaler, E., 1972, *Monatsh. Chem.*, **103**, 1271.
191 Kapfer, E., Schauenstein, E., 1972, *Z. Krebsforsch.*, **77**, 180.
192 Zollner, H., Schauenstein, E., 1973, *Z. Krebsforsch.*, **79**, 108.
193 Scaife, J. F., 1970, *Naturwissenschaften*, **57**, 250.
194 Schauenstein, E., Wünschmann, B., Esterbauer, H., 1969, *Z. Krebsforsch.*, **72**, 325.
195 Schauenstein, E., Ernet, M., Esterbauer, H., Zollner, H., 1971, *Z. Krebsforsch.*, **75**, 90.
196 Schauenstein, E., Zollner, H., Ernet, M., Esterbauer, H., 1971, *Z. Krebsforsch.*, **76**, 146.
197 Krings, R., Tritsch, H., 1973, *Arch. Dermatol. Forsch.*, **246**, 108.
198 Musger, A., 1969, *Wien. Med. Wochenschr.*, **6**, 117.
198a Tillian, H. M., Schauenstein, E., Ertl, A, Esterbauer, H., 1977, *Europ. J. Cancer* (in press).

198b Conroy, P. J., Nodes, J. T., Slater, T. F., White, G. W., *Europ. J. Cancer* (in preparation).
199 Musger, A., 1969, *Verh. Dermatol. Ges.*, **20**, 375.
200 Musger, A., 1968, *Lecture: 4. Dermatol. Colloqu. Bühl.*
201 Ratzenhofer, M., Richter, K., Schauenstein, E., 1975, *Exp. Pathol.*, **11**, 83.
202 Dostal, V., Schauenstein, E., Kulnigg, P., Schmeller, E., 1974, *Z. Naturforsch.*, **29c**, 76.
203 Kulnigg, P., 1973, Dissertation, University of Graz.
204 Burkl, W., Klingenberg, H., Schauenstein, E., Taufer, M., 1968, *Wien. Klin. Wochenschr.*, **80**, 238.
204a Esterbauer, H., Sanders, E. B., Schubert, J., 1975, *Carbohydrate Res.*, **44**, 126.
205 Booth, J., Boyland, E., Sims, P., 1961, *Biochem. J.*, **79**, 576.
206 Al-Kassab, S., Boyland, E., Williams, K., 1963, *Biochem. J.*, **87**, 4.
207 Grover, P., Sims, P., 1964, *Biochem. J.*, **90**, 603.
208 Boyland, E., Williams, K., 1965, *Biochem. J.*, **94**, 190.
209 Johnson, M., 1966, *Biochem. J.*, **98**, 44.
210 Boyland, E., Chasseaud, E., 1967, *Biochem. J.*, **104**, 95.
211 Boyland, E., Chasseaud, E., 1968, *Biochem. J.*, **109**, 651.
212 Speir, T. W., Barnsley, E. A., 1971, *Biochem. J.*, **125**, 267.
213 Williams, R. T., 1959, *Detoxification Mechanisms*, 2nd edn. (John Wiley, New York), p.336.
214 Fischer, F. G., Bielig, H.-J., 1940, *Z. Phys. Chem.*, **266**, 73.
215 Fischer, F. G., Wiedemann, O., 1936, *Ann. Chem.*, **520**, 52.
216 Hardinge, M. G., Frazer, J. M., Hahn, I. J., 1963, *Fed. Amer. Soc. Exp. Biol.*, **22**, 367.
217 Frazer, J. M., Peters, M. A., Helgar, H., Hardinge, M. G., 1965, *Fed. Amer. Soc. Exp. Biol.*, **24**, 426; 1966, **25**, 557.
218 Frazer, J. M., Peters, M. A., Hardinge, M. G., 1967, *Mol. Pharmacol.*, **3**, 247.
219 Esterbauer, H., Scholz, N., Sterk, H., 1972, *Monatsh. Chem.*, **103**, 1453.
220 Binder, M., 1969, Dissertation, University of Graz.
221 Sund, H., Theorell, H., 1963, *The Enzymes*, volume 7, page 25 (Academic Press, New York).

α-Hydroxyaldehydes

4.1 Glyceraldehyde

Over forty years ago Mendel reported that L-glyceraldehyde, but not D-glyceraldehyde, inhibited enzyme activity of the Jensen sarcoma *in vitro*[1]. In this connection the results of Meyerhof et al.[2] should be mentioned, and also those of Embden et al.[3], showing that L-glyceraldehyde condenses with hydroxyacetone phosphate to form sorbose 1-phosphate which, according to Lardy et al.[4], inhibits hexokinase. Lardy et al. explain the inhibitory action of L-glyceraldehyde on glycolysis in this way.

Since then many relevant investigations on the action of D- and L-glyceraldehydes on intermediary metabolism have appeared, but only the more recent investigations which have shed new light on the topic will be mentioned here.

In general the fact emerges that high concentrations of glyceraldehyde are needed to produce a medium-to-strong inhibition. It should be recalled that the aldehyde in its L-configuration appears in normal metabolism as a metabolite. The concentrations required for effective inhibition lie in the region of 10 mM. Thus glyceraldehyde cannot be said to have a particularly strong inhibitory effect on the energy metabolism.

In 1964 a distinct inhibition of the hexokinase of EATC was found *in vitro* for 10 mM glyceraldehyde. D,L- and L-glyceraldehyde effect a complete loss of activity, whereas D-glyceraldehyde produces a significant but incomplete loss of activity[5]. It was nevertheless shown that Lardy's explanation about the selective activity of L-glyceraldehyde does not apply generally or unconditionally.

Spolter et al.[6] have reported on the inhibitory effect of different aldehydes that possess carbonyl or hydroxyl groups vicinal to the aldehyde function, on the system aldolase + fructose 1-phosphate. Different inhibitory effects were shown by glyceraldehyde on aldolases from rabbit muscle or liver. Simple aldehydes were shown to be inactive.

The investigations were further extended to aerobic metabolism as well as to the biosyntheses of nucleic acids and proteins. Thus, in 1964 Ölkers et al.[7] examined both stereoisomers of glyceraldehyde for their action on the incorporation of thymidine into DNA, and uracil into RNA, of Ehrlich tumour cells, using both substances in 10 mM solutions aerobically and anaerobically. It was shown that the D-form, used aerobically, inhibits thymidine incorporation by 80%–90%, but L-glyceraldehyde only to 50%. Under anaerobic conditions D-glyceraldehyde produces practically no inhibition, whereas L-glyceraldehyde produces nearly 100% inhibition. In comparison, RNA synthesis is distinctly more resistant: under aerobic conditions the inhibitions lie between 23% and 28%. Anaerobic action again leaves the effect of D-glyceraldehyde unchanged (approximately 20%), but the inhibition effected by L-glyceraldehyde rises practically to 100%.

Preincubation of the cells with D-glyceraldehyde gives a much greater inhibition (70%) than with L-glyceraldehyde (20%).

In 1967 Langen et al.[8] examined the action of D,L-glyceraldehyde in 10 mM solution *in vitro* on biosyntheses of DNA, RNA, and protein in Ehrlich cells and in a cell-free system from mouse liver. In agreement with the findings of Ölkers et al.[7], strong inhibition was detected for the nucleic acid biosyntheses. The observation that protein synthesis is clearly more susceptible, being inhibited even by 0·1 mM solutions, is of decisive importance. The mechanism of inhibition of protein synthesis was examined in more detail for the cell-free system. In this case the incorporation of phenylalanine into ribosomal peptides is inhibited only slightly by D,L-glyceraldehyde in 10 mM solution, if the ribosomes are already loaded with endogenous messenger RNA or with poly-U. If D,L-glyceraldehyde is used at a time when the ribosomes still have to be loaded with poly-U, inhibition of peptide synthesis results. This effect may be explained if D,L-glyceraldehyde prevents the linking of messenger RNA to the ribosomes. In model experiments it was shown that D,L-glyceraldehyde undergoes reaction directly with poly-A, poly-U, and poly-A–poly-U. The action of D,L-glyceraldehyde on tumour cells might therefore consist of an instantaneous action on the information transfer system, probably by direct reaction with single-stranded nucleic acids. The strong inhibitory action on protein synthesis precludes this substance, in the authors' opinion, being recommended for use as an anticancer drug.

The investigations of Guidotti et al.[9] bring forward a new aspect concerning the mechanism of action of the aldehyde in intermediary metabolism: glyceraldehyde, like other aldehydes examined, inhibits protein biosynthesis and the glycolysis of Yoshida ascites hepatoma cells *in vitro*. This inhibition is accompanied by a decrease in the amount of intracellular cysteine. If cysteine is added before allowing the aldehyde to act upon the system, the inhibition is prevented, whereas later addition of cysteine partially reverses the inhibition. The decrease in the intracellular concentration of sulphydryl compounds is explained by the formation of thiazolidinecarboxylic acids, which is seen as the first step in the development of the observed inhibitory effects. Two years after this work Markgraf-Modersohn et al.[10] examined the action of thiazolidinecarboxylic acids from L-glyceraldehyde and cysteine, or of penicillamine, on the glycolysis and respiration of Ehrlich tumour cells *in vitro*. It was found that these compounds can inhibit only aerobic glycolysis and that to a maximum of 40%–50%. Anaerobic glycolysis and respiration remain unaffected. Since free L-glyceraldehyde, as had been shown by Mendel[1] and later by Warburg et al.[22], inhibits anaerobic glycolysis almost completely, Markgraf-Modersohn et al. concluded that the thiazolidines of L-glyceraldehyde can generate free glyceraldehyde in the cell only under aerobic conditions, and this then inhibits the glycolysis. An inhibition of respiration by the thiazolidines was not even expected, since free L-glyceraldehyde was not able to inhibit measurably the respiration under the same experimental conditions. Clearly under anaerobic conditions significant cleavage of the thiazolidines does not take place. The thiazolidinecarboxylic acids of L-glyceraldehyde proved to be such weak inhibitors *in vitro* that the authors did not proceed with experiments on the living animal.

Fasske and Morgenroth[11] are concerned with cytomorphological investigations on the effect, after intratumoral injection *in vivo*, of glyceraldehyde on epithelial tumours caused by 9,12-dimethyl 1,2-dibenzanthracene. Electron microscopic investigations of the subsequently excised tumours revealed, as a consequence of the expected primary inhibition of hexokinase[4], deformations of the endoplasmatic reticulum which, however, still bore ribosomes. On further action of the aldehyde the mitochondria swell, and the activity of succinate dehydrogenase is no longer detectable. At the same time the ribosomes shrink. These effects were interpreted as resulting from an inhibition of the citric acid cycle and a blocking of protein biosynthesis. Disordered dehydrations and cross-linking of the cytoplasm finally lead to cell death. This work is especially informative, since it correlates previous knowledge about the inhibitions of the energy metabolism and the biosyntheses of nucleic acids and proteins *in vitro* with enzyme–histochemical evidence and morphological changes *in vivo*.

In their investigations on the influence of D,L-glyceraldehyde on glucose metabolism, Riddick and Bressler[12] reported that the steady-state level of ATP is lowered, whereas that of glucose 6-phosphate is raised. Wand and Bacigalupo[13] compared the inhibitory effects of D,L- and D-glyceraldehyde in solutions of 1–100 mM on the energy metabolism of Walker carcinoma cells and rat liver cells *in vitro*. The effects of the D-form on various metabolic processes occurring in the mitochondria of both cell types are presented in table 4.1.

Table 4.1. Effects of D-glyceraldehyde on metabolic processes in the mitochondria of liver and Walker carcinoma cells (according to Wand and Bacigalupo[13]).

Process	Liver	Walker carcinoma
Glucose oxidation	no effect	small inhibition
P requirement	unaltered	distinctly decreased
Oxidative phosphorylation	coupled	uncoupled
Succinate oxidation	decreased	more strongly decreased
P requirement	decreased	more strongly decreased
Pyruvate oxidation	no effect	appreciably inhibited
Phosphorylation	no effect	appreciably inhibited
P/O ratio	no effect	appreciably inhibited
Swelling of the mitochondria	none	swelling occurs

It seems remarkable that glyceraldehyde shows different activities with tumour cells and with liver cells, and that the D-form is active. This shows that metabolic blocking may not be called forth solely by the L-form (cf Mendel[1] and Warburg et al.[22]). It is clear that such manifold and drastic intervention in the energy metabolism and the biosynthetic processes must lead to inhibition of growth and eventually to the death of the cell. In fact, numerous authors and research groups have investigated this action of glyceraldehyde on bacteria, viruses, and tumour cells. Thus Cutinelli and Caldiero[14] have reported on the partial inhibition of the growth of *E. coli*, and they consider that glyceraldehyde, used aerobically as the racemate, binds cysteine and thereby causes a shortage of cysteine. Glucose utilization and the incorporation of leucine are not measurably affected. Bacterial growth is inhibited more strongly by anaerobic action of D,L-glyceraldehyde. Glycolysis and the incorporation of phosphate and ^{14}C fragments are also partially inhibited under these conditions.

From the results of Ölkers et al.[7] it may be concluded that the biosyntheses of DNA and RNA are also inhibited. For EATC these were inhibited to a considerably greater extent under anaerobic rather than under aerobic action of L-glyceraldehyde. To explain the inhibition of growth, Cutinelli et al. again invoke the binding of cysteine, and also partial inhibition of glycolysis.

Carrera[15] examined the action of D,L-glyceraldehyde on the growth and chromogenesis of *Pseudomonas aeruginosa*, and established that inhibition of growth was dependent on the nature of the medium used. The formation of fluorescein proved to be independent of cell growth.

Mücke and Sproessig[16] described, as part of their investigations on the inhibition of the influenza virus by α-carbonyl compounds, the considerable deactivation *in vitro* of the virus by glyceraldehyde. They found that for dicarbonyl compounds to be active against the influenza virus a vicinal arrangement of the carbonyl groups is required. Substances are therefore inhibitors, if they already possess the α-dicarbonyl grouping

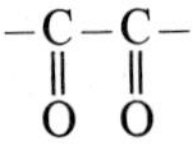

$$-\underset{\underset{O}{\|}}{C}-\underset{\underset{O}{\|}}{C}-$$

as such, or if this grouping may be formed spontaneously or in enzyme-catalysed rearrangements. Acetylacetone is thus inactive, whereas glyceraldehyde is active, since here the possibility exists of formation of methylglyoxal, according to the reaction shown in scheme 4.1.

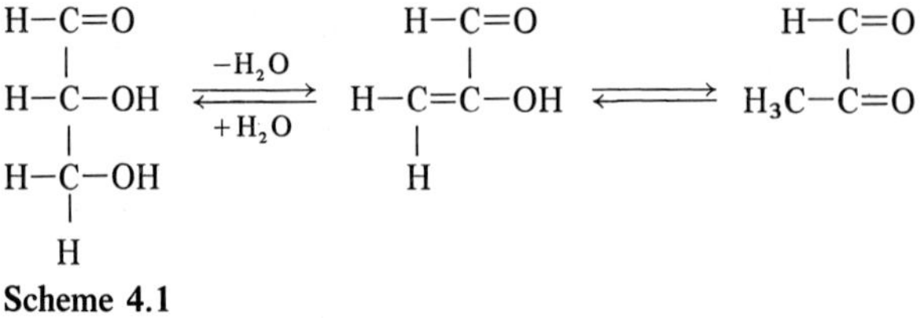

$$\begin{array}{ccccc} H-C=O & & H-C=O & & H-C=O \\ | & \xrightarrow{-H_2O} & | & & | \\ H-C-OH & \xleftarrow{+H_2O} & H-C=C-OH & \rightleftarrows & H_3C-C=O \\ | & & | & & \\ H-C-OH & & H & & \\ | & & & & \\ H & & & & \end{array}$$

Scheme 4.1

It could be established experimentally that this reaction actually takes place. Riddle and Lorenz[17] found that the conversion of glyceraldehyde into methylglyoxal could be catalysed at pH 7·4 and 40°C by Tris buffer, arsenite, arsenate, and phosphate (as inorganic phosphate and as triose or hexose phosphate).

Bonsignore et al.[18] found a factor in bovine liver which catalyses the conversion of glyceraldehyde into methylglyoxal. Subsequently they identified this factor as lysine[19]. The velocity constant of the lysine-catalysed conversion was found to be $1·8 \times 10^{-3}$ ml h^{-1} mol^{-1}, compared with that for the phosphate-catalysed reaction $5-6 \times 10^{-4}$ ml h^{-1} mol^{-1}.

The biochemical and biological significance of these results appears considerable. For glyceraldehyde the possibility arises that the metabolic inhibitions of tumour growth already described, as well as others to be described below, may actually be due to methylglyoxal. Phosphate is present in the suspension media commonly employed, in concentrations equivalent to those in cells and tissues, where lysine will also be present. This role of methylglyoxal is particularly likely since remarkably high concentrations of glyceraldehyde are required to achieve the more powerful inhibitory effects, and since α-ketoaldehydes (as described in detail in section 5) are potent inhibitors of metabolism and of tumour growth. Section 5.4 in particular will examine the significance of the results relating to the presence of methylglyoxal or its derivatives in normal cells and tissues.

About thirty years ago the first data appeared in the literature on the inhibition of growth of animal tumour cells by glyceraldehyde. Riley and Pettigrew[20] reported a small retardation in the appearance and growth of benzopyrene-induced sarcomata, when glyceraldehyde was injected subcutaneously in doses of 0·5 ml of a 0·1 M solution twice weekly for a period of sixteen weeks.

Twenty years later Sartorelli et al.[21] reported experiments on the chemotherapy of several ascites tumours with glyceraldehyde. They found that intraperitoneal injections of D,L-glyceraldehyde extended the survival period of mice with sarcoma-180 or hepatoma-134. Mecca lymphosarcoma proved resistant, however. The combination of glyceraldehyde with purine derivatives (6-thioguanine, 6-chloropurine) showed more pronounced effects than the components used separately.

In 1963 a paper by Warburg et al.[22] appeared with the sensational title 'Heilung von Mäuse-Asciteskrebs durch D- und L-Glycerinaldehyde' ('Cure of ascites cancer in mice by D- and L-glyceraldehyde'). In the historical introduction to this paper only the unpublished and evidently unsuccessful therapeutic experiments of Schoeller were cited, and it was presumed that up to that time only the systemic application of the substance had been attempted, which could not be successful because of the rapid metabolism of the aldehyde. The experiments of Warburg et al. were then reported, which used direct action of the substance on the tumour cells. As the earlier experiments of Riley and Pettigrew[20] and Sartorelli et al.[21] were not mentioned, the impression of a priority was created, which *de facto* did not exist. Warburg *et al.* then proceeded to state that the highly virulent EATC were no longer able to take hold after 30 minutes preincubation *in vitro* in 10 mM solution of L- or D-glyceraldehyde, on reimplantation into the living animal, i.e. they had been killed off. Even when the aldehyde is injected intraperitoneally on the same day or within one or two days of the cell implantation, the cells are killed off to the extent that no animal develops the tumour even after a period of sixty days or more. The rate of killing off amounted to at least 99·998%. The doses required here depend on the time interval between cell implantation and the administration of the injection. Injection on the same day requires 150 μmol/animal; on the second day *post implantationem* 300 μmol/animal is needed. Injections on the third day, even when 400 μmol is used, do not produce a certain cure. The maximal tolerated amount of glyceraldehyde is approximately 1000 mol per day per mouse. The effect on the tumour cells can be achieved only on direct contact, since the compound is metabolized so rapidly when given systemically that it is no longer possible to bring the required concentration of the inhibitor to the tumour cells.

It is understandable that the report by Warburg called forth lively reactions in professional circles, ranging from extreme optimism to absolute rejection. Only one day after the appearance of the Warburg communication Ardenne[23] produced a relevant commentary entitled 'Ein Durchbruch der Chemotherapie des Krebses?' (A breakthrough in the chemotherapy of cancer?'). This was followed by a description of numerous investigations carried out at the Ardenne Institute. These were concerned with the problems of the application, determination of concentrations, theoretical considerations, and numerical estimates relating to the possibilities of maintaining adequately high intravital concentrations of inhibitor.

An opposite view has been represented by several authors whose experimental investigations cast doubt on whether glyceraldehyde is of any significance at all as a cytostatic agent (cf the work of Langen *et al.*[8], discussed above). Two years after the publication of Warburg's communication Brock and Niekamp[24] published detailed results of investigations on the cytostatic action of D-glyceraldehyde on various tumours. Intraperitoneal administration of sublethal doses produced strong inhibitions of Ehrlich ascites tumour as well as of the ascites forms of Yoshida tumour and of DS-carcinosarcoma. The inhibitory effectiveness of D,L-glyceraldehyde with Yoshida ascites hepatoma was confirmed also by Ciaranfi *et al.*[25]. In doses of approximately 100 mg/100 g body weight the substance reduced the number of tumour bearing animals from eighteen to one, when injections were given six days *post implantationem*. α,β-Dihydroxybutyraldehyde at the same dose reduced the number of affected animals from thirty-nine to zero, when administered twenty-four hours *post implantationem*. Motycka and Lako[26] have reported success in curing haemoblastosis L 14 AKR in mice, and in generating cytotoxic effects *in vitro* on leukaemia cells, with glyceraldehyde.

As mentioned several times already all the effects referred to take place only by the direct action of the compound on the tumour cells. In contrast, solid tumours, which in the experiments of Brock and Niekamp[24] caused the death of the animals, are not affected by intravenous, by intraarterial, or even by intratumoral application of the agent, or by regional perfusion. According to Brock and Niekamp D-glyceraldehyde exhibited no specific antitumoral effect and therefore cannot be employed in the chemotherapy of tumours. But in the opinion of Ardenne[27] these results do not contradict theoretical considerations and calculations, or the results obtained *in vitro*.

Gericke[28] likewise examined the activity spectrum of the aldehyde in the mouse, rat, and rabbit and found very diverse activities: complete inhibition of Yoshida ascites hepatoma; negative results with other ascites tumours; no effect, even under the extremely favourable conditions of an intravenous application a few seconds after intravenous inoculation of the tumour. According to Gericke, glyceraldehyde therefore cannot be regarded as a useful cytostatic agent.

Lewerenz[29] found irreversible damage of HeLa cells and FL cells in permanent culture on preincubation in 50 mM and 100 mM solutions of D,L-glyceraldehyde. With fresh, rapidly growing cultures D,L-glyceraldehyde was highly toxic at 10–100-fold lower concentrations.

Hansen and Bohley[30] in 1968 confirmed the previous findings, that intraperitoneal application of D,L-glyceraldehyde in tolerated dosages had no effect on Ehrlich tumour growing intramuscularly. Combination with detergents (octylphenyldecaethylene glycol ether, $0 \cdot 5\%$ in $0 \cdot 9\%$ sodium chloride; laurylamine, $0 \cdot 3\%$; sodium deoxycholate, sodium dodecyl sulphate, sodium oxystearyl sulphate, sodium decylmethyltauride), however, led to an inhibition of the tumour growth.

The investigations of Apple and Greenberg[31] on the inhibition of the growth of animal tumours by 'normal metabolites' are of special interest. They found that glyceraldehyde generally raises the antitumour activity of methylglyoxal. This result is readily understood, considering the direct conversion under physiological conditions of glyceraldehyde into methylglyoxal, as established by Riddle and Lorenz[17] and by Bonsignore *et al.*[19]. According to Apple and Greenberg glyceraldehyde (dose 150–650 mg day^{-1} kg^{-1} for five days intraperitoneally) is more active with Ehrlich ascites carcinoma, sarcoma-180, and adenocarcinoma-755 than is the ketoaldehyde (dose 80 mg day^{-1} kg^{-1}). Methylglyoxal, on the other hand, is more active than glyceraldehyde with leukaemia L 496 and thymolymphosarcoma G C_3HED. Mixtures of both substances are more effective than either separately, and on daily treatment lead to apparently complete remission in 50% of the animals with adenocarcinoma and ascites carcinoma, and in 25% of animals with the sarcoma. A further study was carried out by Giovanella et al.[32] on the action of glyceraldehyde in combination with other substances and other types of experimental procedure. Combination with hyperthermia, often mentioned in connection with tumour therapy, was investigated. D,L-Glyceraldehyde is supposed to act synergistically with hyperthermia, and this treatment appears clinically suitable for man since it also has a low toxicity. The authors base these claims on the examination of leukaemia L 1210 cells, which were studied in a special *in vitro–in vivo* test system.

In 1971 Neish[33] published experiments in which products produced in the reaction of D,L-glyceraldehyde with naturally occurring or synthetic polyamines were examined for antitumour activity. It was shown that the product of the reaction between 4 mol of glyceraldehyde and 1 mol of NN'-bis-(γ-aminopropyl)diaminoethane, when injected intraperitoneally in tolerated dosages, was capable of inhibiting the growth of Rd3-sarcoma, implanted into the right flanks of rats, about 67%. A systemic action has thus clearly been established. However, since the polyamine itself is an effective tumour inhibitor, it cannot be decided whether and to what extent one is dealing with the action of the adduct (possibly also that of the methylglyoxal formed!).

The most recent investigations are those of Sakamoto and Prasad[34], who found that neuroblastoma cells of the mouse show 2·4 times the glycolysis activity, and twice the doubling time, of hamster ovary cells. D,L-Glyceraldehyde (1 mM) decreases glycolysis in the neuroblastoma cells and in the ovary cells to nearly the same extent (approximately 30%). Inhibition of growth in culture amounted to 100% for the neuroblastoma cells and 80% for the ovary cells.

References

1. Mendel, B., 1929, *Klin. Wochenschr.*, **8**, 169.
2. Meyerhof, O., Lohmann, K., Schuster, R., 1936, *Biochem. Z.*, **286**, 319.
3. Embden, G., Schmitz, E., Wittenberg, M., 1913, *Hoppe-Seyler's Z. Physiol. Chem.*, **88**, 210.
4. Lardy, H., Wiebelhaus, V., Mann, K., 1950, *J. Biol. Chem.*, **187**, 325.
5. Drews, J., 1964, *Naturwissenschaften*, **51**, 515.
6. Spalter, P., Adelmann, R., Weinhouse, S., *J. Biol. Chem.*, **240**, 1327.
7. Ölkers, W., Wenzel, M., Schmialek, P., 1965, *Z. Naturforsch.*, **20b**, 227.
8. Langen, P., Bielka, H., Stahl, J., 1967, *Acta Biol. Med. Ger.*, **19**, 17.
9. Guidotti, G., Loreti, L., Ciaranfi, E., 1965, *Eur. J. Cancer*, **1**, 23.
10. Markgraf-Modersohn, D., Siegmund, P., Körber, F., 1970, *Naunyn-Schmiedeberg's Arch. Pharmakol.*, **267**, 241.
11. Fasske, E., Morgenroth, K., 1966, *Arch. Geschwulstforsch.*, **27**, 1.
12. Riddick, J. H., Bressler, R., 1967, *Pharmacology*, **16**, 239.
13. Wand, H., Bacigalupo, G., 1966, *Z. Naturforsch.*, **21b**, 1215.
14. Cutinelle, C., Caldiero, F., 1967, *Boll. Ist. Sieroter. Milan.*, **46**, 418.
15. Carrera, G., 1968, *Riv. Biol.*, 1968, **61**, 99.
16. Mücke, H., Sproessig, M., 1967, *Arch. Exp. Veterinärmed.*, **21**, 307.
17. Riddle, V., Lorenz, F., 1968, *J. Biol. Chem.*, **243**, 2718.

18 Bonsignore, A., Castellani, A., Formaini, G., Segni, P., 1968, *Ital. J. Biochem.*, **17**, 2.
19 Bonsignore, A., Leoncini, G., Siri, A., Ricci, D., 1970, **19**, 284.
20 Riley, J. F., Pettigrew, F., 1944, *Cancer Res.*, **4**, 502.
21 Sartorelli, A., Schoolar, E., Kruse, P., 1960, *Proc. Soc. Exp. Biol. Med.*, **104**, 266.
22 Warburg, O., Gawehn, K., Geissler, A., Lorenz, S., 1963, *Z. Klin. Chem.*, **1**, 175.
23 Ardenne, M. von, 1964, *Dtsch. Gesundheitswes.*, **19**, 709.
24 Brock, N., Niekamp, T., 1965, *Z. Krebsforsch.*, **67**, 93.
25 Cianranfi, E., Loreti, L., Borghetti, A., Guidotti, G., 1965, *Eur. J. Cancer*, **1**, 147.
26 Motycka, K., Lako, L., 1966, *Z. Krebsforsch.*, **68**, 195.
27 Ardenne, M. von, 1965, *Z. Krebsforsch.*, **67**, 230.
28 Gericke, D., 1967, *Abstr. 5th Internat. Congr. Chemother. Vienna*, Part I, 577.
29 Lewerenz, H., 1967, *Z. Krebsforsch.*, **69**, 260.
30 Hansen, H., Bohley, P., 1968, *Z. Krebsforsch.*, **71**, 51.
31 Apple, M., Greenberg, D., 1967, *Cancer Chemother. Rep.*, **51**, 455; **52**, 687 (1968).
32 Giovanella, B., Lehmann, W., Heidelberger, C., 1970, *Cancer Res.*, **30**, 1623.
33 Neish, W. J., 1971, *Z. Krebsforsch.*, **76**, 219.
34 Sakamoto, A., Prasad, K., 1972, *Cancer Res.*, **32**, 532.

Dicarbonyl compounds

5.1 α-Ketoaldehydes
5.1.1 Structures of glyoxal and methylglyoxal

The structure of glyoxal in water was recently investigated by Whipple[1] using NMR spectroscopy. In the early literature the general opinion was that even traces of water effect a polymerization to a high molecular weight polyglyoxal. Whipple could show, however, that in concentrations below 1 M the main species present is a hydrated monomer, and that between 1 M and 10 M two stereoisomeric dimers possessing a five-membered dioxolane ring are chiefly present (cf scheme 5.1).

In agreement with this, the elution pattern on Sephadex G10 shows two peaks with a distribution coefficient $K_d = 0.99$ for the monomer, and $K_d = 2.06$ for the dimer, which is retarded on account of its heterocyclic character[2].

Few investigations on the structure of methylglyoxal in water have been carried out. From the pH dependence of the UV absorption spectrum[3] it may be concluded that at neutral pH a keto–enol equilibrium exists (cf scheme 5.2). The absorption maximum of methylglyoxal at pH 3.4 occurs near 280 nm, and undergoes a bathochromic shift with accompanying increase in intensity with a rise in pH. At pH 10.6 the λ_{max} occurs at 292 nm and can be ascribed to the enolate form. The existence of the enol form in the region of neutrality is of interest because it has the α,β-unsaturated aldehyde structure, which may be responsible for some of the effects caused by methylglyoxal. As discussed in section 3.1, α,β-unsaturated aldehydes have a great affinity for sulphydryl groups, which react at the double bond by a 1,4-addition mechanism. It is possible that the deactivation of sulphydryl enzymes caused by methylglyoxal proceeds according to this mechanism[4,5].

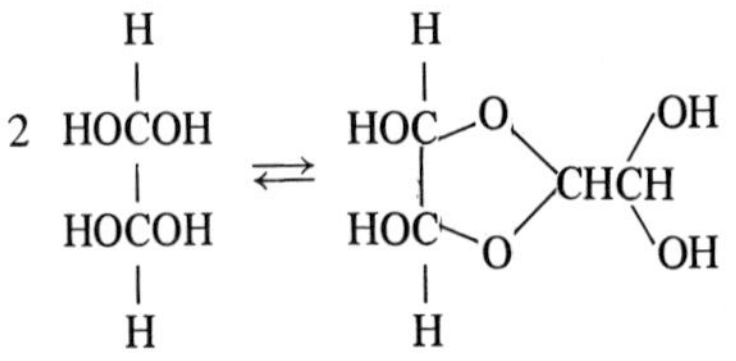

$$2 \begin{array}{c} H \\ | \\ HOCOH \\ | \\ HOCOH \\ | \\ H \end{array} \quad \rightleftarrows \quad \begin{array}{c} H \\ | \\ HOC\diagdown_O \diagup OH \\ \diagdown CHCH \\ HOC\diagup ^O \diagdown OH \\ | \\ H \end{array}$$

monomeric glyoxal dimeric glyoxal

Scheme 5.1

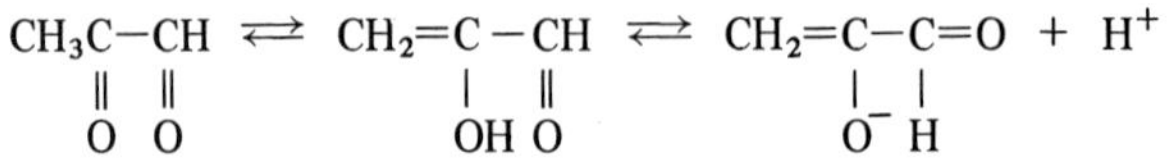

$$CH_3C-CH \rightleftarrows CH_2=C-CH \rightleftarrows CH_2=C-C=O + H^+$$
$$\| \| | \| | |$$
$$O O OH O O^- H$$

Scheme 5.2

5.1.2 Reactions with amines and amino acids

Many compounds of biological importance contain the primary amino group (NH_2), which even in aqueous solution at neutral pH readily undergoes reaction with α-ketoaldehydes. Diverse reaction products can arise in this way, as shown in scheme 5.3. Which of the compounds (I)–(VI) is formed as the main product depends on the kind of amino compound, on the ketoaldehyde, and on the reaction conditions. Königstein and Federonko[6] have shown, using polarographic methods, that the aldehyde group is rapidly converted into the aldimine (II), whereas the keto group reacts only slowly. α-Diketones, such as diacetyl, behave analogously, the first keto group reacting rapidly, the second slowly.

The carbon–nitrogen double bond of the aldimine is about ten to twenty times more stable than the corresponding bond of a ketimine. Consequently, with low concentrations of α-ketoaldehyde, monoamino alcohols (I) and aldemines (II) are the major products. Only at higher concentrations does the keto group react as well to form 1,2-diamino-1,2-dihydroxy compounds (III) and 1,2-di-imines (IV). The reaction products obtained from glyoxal and different aliphatic and aromatic primary amines have been examined by several authors[7-10]. All these reactions are reversible. Compounds (I)–(IV) are not especially stable and decompose again, particularly in acid solutions or even in hot water, into α-ketoaldehydes and amines. Secondary reactions may, however, contribute to appreciable stabilization. Davis and Tabor[11] believe that the protonated compounds of the 1,2-di-imines (cf compound V in scheme 5.3) are particularly stable on account of the symmetry and the possibilities of charge distribution. Maurer and Woltersdorf[12] have demonstrated that 1,2-diamino-1,2-dihydroxyalkanes (III) may furnish stable acid amides (VI) by intramolecular elimination of water. Such stable linkages also probably exist in proteins that have been cross-linked with α-ketoaldehydes.

α-Ketoaldehydes react particularly readily with two amino groups separated by only one or two carbon atoms. With 'methylene diamine', that is a mixture of formaldehyde and ammonia, imidazoles are formed and, with o-phenylene diamine, quinoxalines. The reactions with aromatic o-diamines have also been used for the detection of glyoxals[13]. The α-amino group of amino acids appears to react with α-ketoaldehydes only when a cyclic reaction product can result. Underwood et al.[14] allowed a tenfold molar quantity of different amino acids to act on kethoxal (β-ethoxy-α-oxobutyraldehyde) at pH 7·4. Glycine, lysine, tryptophan, proline, and glutamic acid react hardly at all under these conditions, and free kethoxal is still present after several days. Cysteine, arginine, histidine, serine, and threonine, on the other hand, use up the kethoxal within minutes. The reaction products formed from these amino acids and α-ketoaldehydes have not so far been isolated and characterized; yet, on the basis of the well-known reactions of saturated aldehydes, it may be supposed that cysteine reacts to give thiazolidine-4-

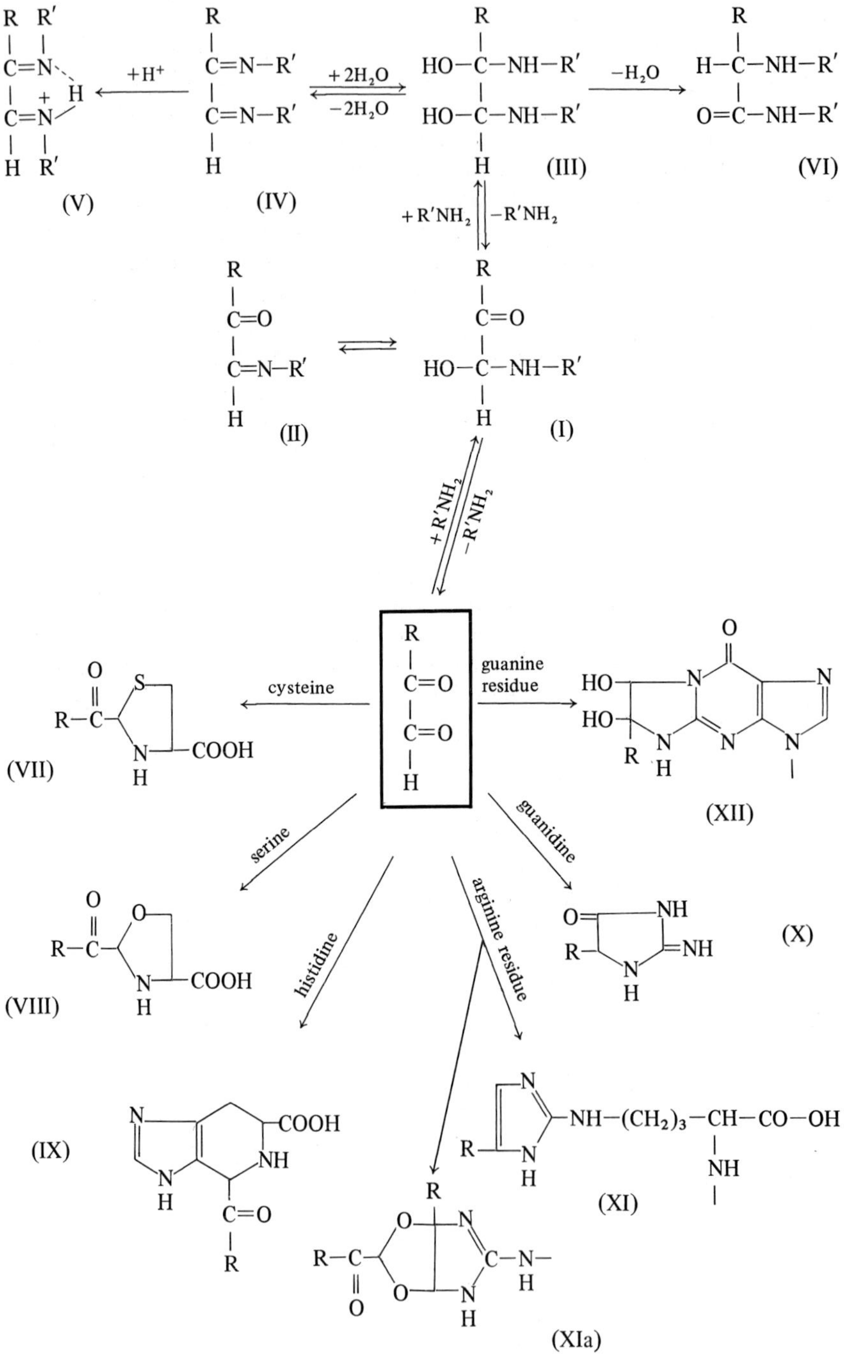
+H+
+2H2O
-2H2O
-H2O
(V)
(IV)
(III)
(VI)
+R'NH2
-R'NH2
(II)
(I)
cysteine
guanine residue
serine
histidine
arginine residue
guanidine
(VII)
(VIII)
(IX)
(X)
(XI)
(XIa)
(XII)

Scheme 5.3

carboxylic acid (VII)[43], serine and threonine to give oxazolidine-4-carboxylic acids (VIII), and histidine to give tetrahydroimidazopyridines (IX). The reaction of cysteine is accompanied by a strong rise in the absorption at 240 nm. According to Bengelsdorf[15] guanidine reacts with glyoxal and methylglyoxal to give glycocyamidines [tautomeric with 5-substituted 2-amino-4-hydroxyimidazoles (X)]. Substituted guanidines, such as arginine, give different reaction products. Bengelsdorf[15] reported that he was unsuccessful in isolating a well-defined reaction product. Bowes and Cater[16] are of the opinion, albeit without any additional evidence, that free arginine or protein-bound arginine reacts with glyoxal or methylglyoxal to give an imidazole derivative (XI). Hydrolysis of the reaction product with concentrated hydrochloric acid gave three degradation products which could be separated on a cation exchanger. Takahashi[17] proposed the structure (XIa) for the reaction product between phenylglyoxal and arginine residues.

It was shown first by Schubert[18], and later by Kermack and Matheson[19] and by Cliffe and Waley[20], that glutathione reacts nonenzymatically with methylglyoxal, and presumably also with other α-ketoaldehydes, to give a hemimercaptal (cf scheme 5.5). The latter is not particularly stable, its dissociation constant being 2 mM at pH 6·6 [20,21]. Higher pH values cause a shift of the equilibrium in favour of free glutathione and methylglyoxal. Formation of the hemimercaptal is measured by the increase in extinction at 240 nm. According to Cliffe and Waley[20], glutathione hemimercaptal is the actual substrate for the glyoxalase reaction.

5.1.3 Reaction with proteins

In principle α-ketoaldehydes can react in proteins with the guanidine group of arginine, the ε-amino group of lysine, and the sulphydryl group of cysteine. However, the products formed differ markedly in their stability. Imidazole derivatives are probably formed with arginine, and these are stable to dialysis, hot water, dilute acids, and alkalies. On hydrolysis of proteins treated with ketoaldehyde several degradation products are obtained from the modified arginine, as shown by Bowes and Cater[16]. With the ε-amino group of lysine ketoaldehydes form Schiff's bases, and with sulphydryl groups presumably hemimercaptals. These reactions are reversible and the products are not stable, readily splitting into their original components. One may achieve reversible decomposition of these Schiff's bases or hemimercaptals either by removing the excess of reagent by means of dialysis or Sephadex filtration, or by treating the protein with acid or alkali. The results show that α-ketoaldehydes block only the arginine residues in proteins irreversibly, whereas the sulphydryl groups and ε-amino groups are blocked reversibly. α-Ketoaldehydes are thus suitable reagents for modifying specifically the arginine residues in proteins. Information can thus be obtained on whether arginine is essential for the structure and function of the protein. For this purpose glyoxal[22] and phenylglyoxal[17] have been particularly recommended. In practice the protein is treated at pH 8–8.6 with an excess of 0.1%–3% glyoxal (or phenylglyoxal) solution. The excess of reagent is removed by dialysis or with Sephadex, and further investigations of the protein (that is, measurement of enzyme activity) may then be undertaken. Nakaya *et al.*[22] have used glyoxal for the specific blocking of arginine residues in lysozyme and insulin. It is interesting that the B_{22}-arginine in normal insulin does not react, although it reacts readily in alkali-denatured insulin or in a pure B-chain. It was concluded that a hydrogen bond exists in normal insulin between B_{22}-arginine and B_{26}-tyrosine. This also shows that the reactivity of arginine residues towards α-ketoaldehydes is substantially influenced by the structure of the peptide chain, and that guanidine groups are less reactive if they are involved in hydrogen bonding.

In asparaginase (*Escherichia coli*) twelve of a total of thirty-three arginine residues can react with glyoxal, and the modified enzyme still has 25% of the original activity[23].

In haemoglobin, of fourteen arginine residues eight react quickly with glyoxal, and two slowly, and the remaining four residues do not react at all. Glyoxal-treated haemoglobin is appreciably more stable towards denaturation than is the untreated haemoglobin; oxygen uptake and release as well as the spectroscopic properties and the sedimentation properties remain unaltered. This is especially of significance as a possible application for the stabilization of stored blood[24]. Takahashi[17,25] has used phenylglyoxal to modify the arginine residues in ribonuclease and structure (XIa) (scheme 5.3) was proposed for the derivative formed. As is shown in figures 5.1 and 5.2, the enzyme activity decreases in parallel with the arginine content. In this case, in addition to arginine, the *N*-terminal amino acid alanine also reacts, but this has no effect on the activity. Keil[26] has shown

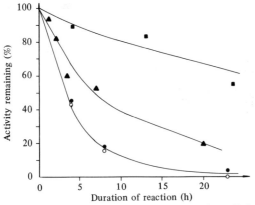

Figure 5.1. Rates of inactivation of ribonuclease T_1 by reaction with phenylglyoxal at pH 8.0 and 25°C. ●, ○, Protein 0.16%, reagent 1.5%; ▲, protein 0.03%, reagent 0.6%; ■, protein 0.09%, reagent 0.15%. Activity was determined towards RNA (●, ▲, ■) and 2′,3′-cyclic guanylic acid (○) (according to Takahashi[25]).

that in pancreas trypsin inhibitor six arginine residues may be converted by means of phenylglyoxal without loss of activity. Glyoxal was also used for the modification of arginine residues in D-amino-acid oxidase[27] and in glycogen phosphorylase[28]. The phosphorylase is completely deactivated within one hour by approximately 1 mM glyoxal at pH 7·9. The reaction velocity is pH-dependent, and increases tenfold with each pH unit. In a review article, Glazer[29] has discussed the use of 1,2-dicarbonyl compounds for the specific modification of arginine residues in proteins.

When α-ketoaldehydes, in particular glyoxal, act in high concentrations (about 10%) on proteins, stable inter- and intramolecular cross-linkages may also develop, as shown for gelatin[11,30], gliadin[31,32], and collagen[33]. Davis and Tabor[11] believe that these bridgings between amino groups are based on a protonated complex; a second possibility is the production of a peptide linkage[12] (scheme 5.4).

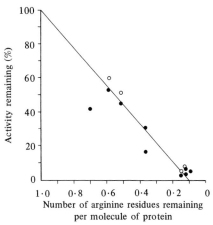

Figure 5.2. Relationship between loss of arginine residues and extent of inactivation of ribonuclease T_1 by reaction with phenylglyoxal at pH 8·0 and 25°C. Activity was determined towards RNA (●) and 2',3'-cyclic guanylic acid (○) (according to Takahashi[25]).

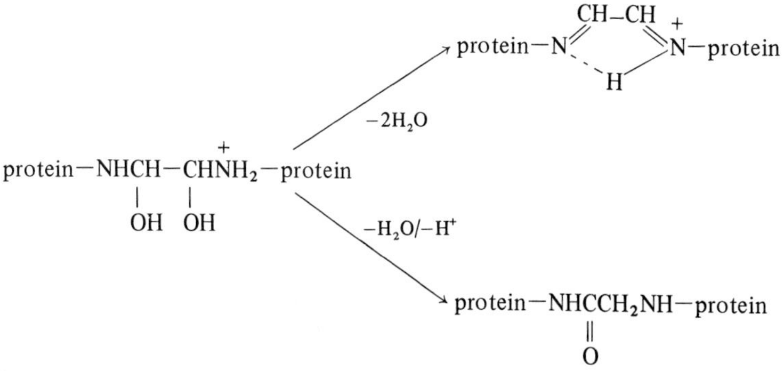

$$\text{protein}-\overset{+}{\text{NHCH}}-\text{CHNH}_2-\text{protein} \quad \overset{-2H_2O}{\longrightarrow} \quad \text{protein}-N\overset{CH-CH}{\underset{H}{\diagup \quad \diagdown}}\overset{+}{N}-\text{protein}$$

$$\underset{\text{OH} \quad \text{OH}}{} \quad \overset{-H_2O/-H^+}{\longrightarrow} \quad \text{protein}-\text{NHCCH}_2\text{NH}-\text{protein}$$
$$\underset{O}{\parallel}$$

Scheme 5.4

5.1.4 Reactions with nucleic acids and their building blocks

The possible significance of α-ketoaldehydes as antiviral or carcinostatic compounds has prompted detailed investigations of the reactivity of such compounds towards nucleic acids and their building blocks.

Staehelin[34] has shown that glyoxal and kethoxal, like formaldehyde, destroy the infectivity of tobacco mosaic virus DNA. Ten to fifteen molecules of glyoxal or kethoxal are bound to the isolated nucleic acid of a virus particle under conditions which lead to 50% deactivation, whereas about one thousand to two thousand molecules are bound to the complete virus particle. This high number of bound molecules clearly relates to the protein coat, whose arginine residues may be bound to glyoxal. Deactivation, however, proceeds even with the intact virus by direct reaction of the ketoaldehyde with the virus nucleic acid. Staehelin drew the conclusion, based on UV spectroscopic investigations, that glyoxal and kethoxal, in contrast to formaldehyde, react specifically with guanine bases. He did not succeed in isolating well-defined adducts of guanine derivatives with glyoxal or kethoxal, but he did obtain a well-defined 1:1 adduct of isocytosine with kethoxal. Staehelin has proposed formula (XII) in scheme 5.3 for the guanine–α-ketoaldehyde adduct. This structure was confirmed by subsequent investigations carried out by Shapiro and Hachmann[35], and Shapiro et al.[36]. These authors have obtained the adducts of guanine and guanosine with glyoxal, methylglyoxal, and kethoxal, and determined their structures by NMR, IR, and UV spectroscopy, and elementary analyses. The adducts are not particularly stable and rapidly dissociate to the starting products in weakly alkaline solutions. Nakaya et al.[37] examined the reactions of glyoxal with nucleic acids, nucleotides, and their bases, by measurement of the change in UV light absorption. The spectra of all the bases and nucleotides are altered in the presence of large amounts of glyoxal (approximately 150 mM). For low glyoxal concentrations, however, only guanine and guanosine monophosphate produce a characteristic spectral change. The reaction is complete within 30 minutes at pH $8 \cdot 3 - 8 \cdot 9$, and may be followed by measurement of the decrease of the UV absorption at 269 nm. Figures 5.3 and 5.4 show the dependence of the extent of the reaction on the pH and the glyoxal concentration respectively. The formation of adduct is probably an equilibrium reaction and an approximately two hundredfold excess of glyoxal is required to effect complete conversion (cf figure 5.4). Nakaya et al.[37] have shown that the other nucleic acid bases (cytosine, adenine, uracil) also undergo equilibrium reactions when very high glyoxal concentrations are employed. Such adducts are, however, so unstable that dissociation to the starting substances takes place when the solution is merely diluted. The specificity of α-ketoaldehydes towards guanine is consequently caused by the fact that these adducts are substantially more stable than the adducts with other bases. Because of this specificity, glyoxal and kethoxal are more suitable than other reagents for the determination, in nucleic acids, of the number of guanine residues which are accessible and not involved in hydrogen bonds.

Nakaya et al.[37] also found that normal calf thymus DNA gave no reaction with glyoxal or kethoxal because all the guanine residues are blocked by the hydrogen-bonded double strand. Heat-denatured DNA, however, does react with glyoxal or kethoxal, albeit more slowly than with the free bases or the nucleotides. Litt and Hancock[38] and Litt[39] have used kethoxal for the determination of the nucleotide sequence in the single-stranded region of t-RNA. Other authors[40] have used glyoxal for the same purpose.

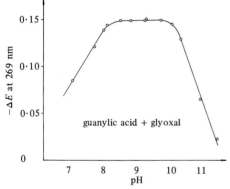

Figure 5.3. Effect of pH on the degree of reaction of 47·8 μM guanylic acid by 4·0 mM glyoxal after 2 h of incubation, as estimated from the decrease in extinction at 269 nm (according to Nakaya et al.[37]).

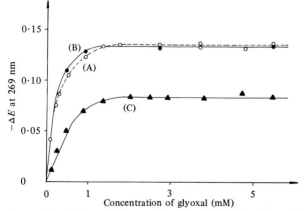

Figure 5.4. Values of decrease in extinction obtained at pH 8·9 for 42·4 μM guanylic acid (A), an equimolar mixture of 42·4 μM nucleotides (B) and 65·1 μM guanine (C) plotted as a function of glyoxal concentration (according to Nakaya et al.[37]).

Guanine residues in the single-stranded region react appreciably more rapidly with kethoxal and glyoxal than in the double-stranded region. It is of significance for these investigations that kethoxal is also obtainable in the tritium- or ^{14}C-labelled forms.

α-Ketoaldehydes are able to react with the DNA of intact cells only when the DNA is uncoiled, the guanine bases thus becoming accessible, that is, during replication and transcription. Experiments by Brooks and Klamerth[41] showed that glyoxal is also able to undergo reversible interaction with DNA in intact cells. DNA isolated from glyoxal-treated fibroblasts shows an altered melting-point diagram, altered chromatographic behaviour, and is incompletely degraded by deoxyribonuclease. Klamerth[42] has also found alterations in the RNA from glyoxal-treated fibroblasts.

5.1.5 Effects on different cell functions
5.1.5.1 *Uptake of oxygen*
The effects of methylglyoxal, and glyoxal in particular, have been closely examined. These aldehydes may act on respiration both as inhibitors and as stimulators. Which effect operates depends on the nature of the aldehyde and the tissue. It is known that α-ketoaldehydes may be converted into α-ketoacids by α-ketoaldehyde dehydrogenases, and to α-hydroxy acids by the glyoxalase system. The activity of the two enzyme systems in the tissues or in the isolated mitochondria will essentially determine whether the result will be an inhibition, or a stimulation, of respiration.

In 1932 Kisch[43] gave the first report that methylglyoxal inhibits the respiration *in vitro* of kidney, retina, diaphragm, and liver of rats and cattle, as well as of Jensen sarcoma of the rat. The inhibition of respiration is specially marked for the kidney tissues. Cardiac tissue occupies a special position of interest, as its oxygen consumption is not inhibited by methylglyoxal but is increased by 30%-100%. Further investigations by Kuhn[44] have shown that methylglyoxal deactivates functional sulphydryl group enzymes which participate in energy metabolism. It was assumed that methylglyoxal forms stable complexes directly with the sulphydryl groups and probably also with the amino groups of the enzymes.

Methylglyoxal inhibits the following individual enzymes: hexokinase, aldolase, and glyceraldehyde 3-phosphate, glutamate, malate, and succinate dehydrogenases, as well as adenosine triphosphatase. Spolter *et al.*[45] examined in detail the inhibition of aldolase, and Kuhn[44] that of succinate dehydrogenase. In isolated rat liver mitochondria the inhibition of SDH runs parallel to the loss of mitochondrial sulphydryl groups. The inhibition is irreversible and cannot be abolished either by repeated washing or by means of glutathione. However, if glutathione is given simultaneously with methylglyoxal it has a protective effect, and with increasing concentration prevents the inhibition, since it reacts with methylglyoxal to give an inactive product (namely thiohemiacetal; cf scheme 5.5).

Similarly to methylglyoxal, glyoxal inhibits enzymes for whose activity sulphydryl groups are required. Spolter et al.[45] have again reported on the inhibition of aldolase, and Kuhn[46] on the inhibition of hexokinase and triose phosphate dehydrogenase. In this paper Kuhn also reports that glyoxal blocks the respiration of heart, kidney, and brain slices of the rat, but stimulates the respiration of liver slices, in contrast to methylglyoxal. The same observation was made by Horton and Packer[47] on isolated liver mitochondria. Glyoxal has no influence on the respiratory capacity of cells in cultures of human fibroblasts[48].

Glyoxal inhibits oxygen uptake, and aerobic and anaerobic glycolyses, in Yoshida ascites hepatoma cells[49]. It is also significant that glyoxalate inhibits the respiration of many tissues by blocking aconitase[50]. Glyoxalate is formed from glyoxal by the action of an α-ketoaldehyde dehydrogenase. It is therefore conceivable that in some cases it is not glyoxal itself, but glyoxalate, that is the actual inhibitor.

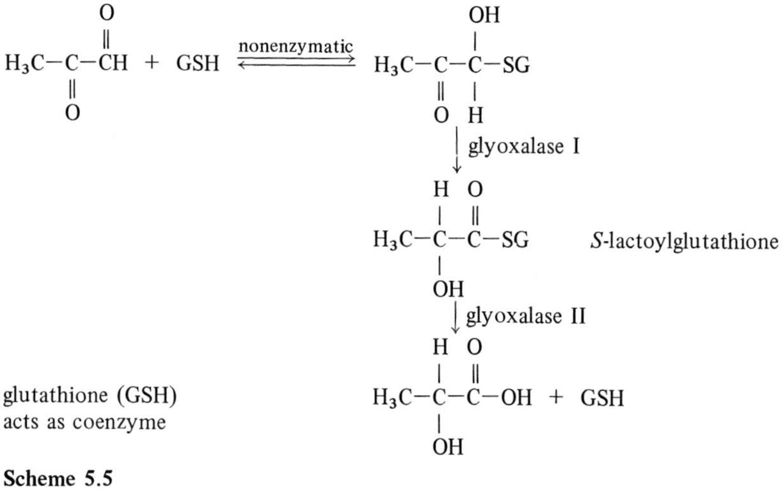

$$\underset{\substack{\|\\O}}{H_3C-\overset{\overset{\displaystyle O}{\|}}{C}-CH} + GSH \; \underset{}{\overset{nonenzymatic}{\rightleftarrows}} \; H_3C-\underset{\underset{\displaystyle O}{\|}}{\overset{\overset{\displaystyle OH}{|}}{C}}-\underset{H}{\overset{|}{C}}-SG$$

$$\downarrow \text{glyoxalase I}$$

$$H_3C-\underset{\underset{\displaystyle OH}{|}}{\overset{\overset{\displaystyle H}{|}}{C}}-\overset{\overset{\displaystyle O}{\|}}{C}-SG \qquad S\text{-lactoylglutathione}$$

$$\downarrow \text{glyoxalase II}$$

$$H_3C-\underset{\underset{\displaystyle OH}{|}}{\overset{\overset{\displaystyle H}{|}}{C}}-\overset{\overset{\displaystyle O}{\|}}{C}-OH + GSH$$

glutathione (GSH) acts as coenzyme

Scheme 5.5

5.1.5.2 α-Ketoaldehydes as regulators and inhibitors of cell growth
α-Ketoaldehydes are widely known as potent inhibitors of the proliferation of bacterial, plant, and animal cells. Együd, Szent-Györgyi, and others have examined the inhibitory effects of α-ketoaldehydes on cell proliferation in considering aspects of a special theory. Their hypothesis is that in multicellular organisms a universal mechanism regulates growth (that is, cell division). The same factor, which permanently promotes growth, one of the main characteristics of life, would exist in all living systems and would be controlled in its action by an antagonist. The effect which finally resulted would be based on the equilibrium between the two factors.

The many different changes experienced by a cell during division would all be regulated by the same receptor, which would receive the signals of the promotion factor and the inhibitory factor. This receptor would involve the sulphydryl group, whose indispensability for cell division has long been known.

The inhibitory factor, known as 'retine', was considered to be a glyoxal derivative, since certain tissue extracts which had growth-inhibiting activity had been obtained several years earlier. After partial purification IR spectroscopic evidence of the presence of α-ketoaldehydes[51,52] was obtained.

According to Együd and Szent-Györgyi[53,54], glyoxal derivatives with chain lengths of from three to thirteen carbon atoms, with the exception of the twelve carbon atom homologue, strongly inhibit the growth of *Escherichia coli* when used in 1 mM solutions. Methylglyoxal in 1 mM solution brings to a standstill the growth not only of *Escherichia coli*, but also of flagellates, germinating seeds, fertilized sea urchin eggs, and tissue cultures, without apparent damage to the cells. Further, it has long been known that carbonyl compounds react readily with thiol groups[18,55,58].

The assumption that the growth-inhibiting action of 'retine' depends on a blocking of sulphydryl groups, which are functional for growth, thus appears plausible.

The accelerating factor, termed 'promine', is considered to involve the glyoxalase enzyme system, which is known to convert α-ketoaldehydes into the biologically much less reactive α-hydroxycarboxylic acids (cf scheme 5.5).

The inhibitions of growth produced by the glyoxal derivatives mentioned are reversible; their action *in vitro* only lasts for several hours. Addition of sulphydryl compounds, such as cysteine or cysteamine, cancels the inhibitory action instantaneously[56,57]. This effect may be explained by the ideas referred to above.

The inhibitory action of α-ketoaldehydes is less dependent on the concentration of the solution of the inhibitor than it is on the ratio of the quantity of aldehyde used to the number of cells. This result suggests that a stoichiometric reaction exists between the aldehyde and certain sulphydryl groups which are clearly essential for cell proliferation.

The question remains as to which particular type of cellular material these sulphydryl groups belong. Cell respiration may be excluded as this system for practical purposes, since it was shown that 1 mM solutions of methylglyoxal, which stop cell proliferation, do not measurably influence the oxygen uptake of *Escherichia coli*[53] or of mammalian cells[48].

It has further been shown that 1 mM solutions of methylglyoxal block protein biosynthesis in *Escherichia coli* to the extent of 96%, whereas biosyntheses of DNA and RNA continue at a 50% or 70% level respectively, although growth has ceased[59].

Scaife[57] treated synchronized kidney cells with kethoxal and methylglyoxal and found that both compounds in concentrations of approximately 100 μg/ml strongly inhibit protein synthesis in the S-phase as well as in the G_2-phase, and that RNA and DNA syntheses were influenced to a smaller extent. Otsuka and Együd[60] also found in the cells of sarcoma-180 that methylglyoxal inhibits virtually only the synthesis of proteins, and not that of RNA. They therefore believe, as does Greeg[61], that α-ketoaldehydes quite generally inhibit protein synthesis and thereby block cell proliferation. It may be supposed that the points of attack for the α-ketoaldehydes in the inhibition of cell proliferation are the functional sulphydryl groups involved in protein biosynthesis.

Szent-Györgyi et al.[62,63] and Knock[64] believe that sulphydryl-rich residual proteins, which partly surround the DNA and contribute to the regulation of cellular activity, become modified by α-ketoaldehydes.

Guidotti et al.[49] concluded that it is the high affinity of glyoxal for cysteine is responsible for the inhibition of protein synthesis. Similarly to other aldehydes, glyoxal condenses with cysteine to give stable thiazolidinecarboxylic acids, which results in decrease in intracellular cysteine concentration and a disturbance of the amino acid pool necessary for protein synthesis. Glutathione, however, forms only thiohemiacetals, which are readily dissociable; glutathione is therefore unable to deactivate the aldehyde.

α-Ketoaldehydes (glyoxal) and α-hydroxyaldehydes (glyceraldehyde, glycolaldehyde) inhibit the leucyl-t-RNA synthetase of rat liver, and hence may also influence protein synthesis in this way[65].

Klamerth[48] has shown that glyoxal causes a general inhibition of cell growth, and of the synthesis of DNA, RNA, and proteins, in cultures of human fibroblasts. The incorporation of amino acids into protein is not inhibited completely even after prolonged action, but DNA synthesis is completely blocked. Syntheses of m-RNA and ribosomal RNA are inhibited appreciably less strongly than is DNA synthesis. Glyoxal has no influence *in vitro* on thymidine kinase and on DNA-dependent RNA polymerase. In contrast to the other authors mentioned above, Klamerth comes to the conclusion that the growth-inhibiting action of glyoxal and other α-ketoaldehydes depends primarily on the inhibition of DNA replication. It was assumed that α-ketoaldehydes react with the protein coat (arginine residues) of DNA and thereby prevent separation of the strands; or, alternatively, that these aldehydes modify the guanine residues of DNA directly and in particular during the replication process, when the two strands are separated and the bases are therefore more reactive. In favour of a direct effect on the DNA is the observation that DNA obtained from aldehyde-treated fibroblasts shows characteristic changes, namely in its temperature absorption profile, chromatographic behaviour, and degradation by deoxyribonuclease. In hamster cell cultures chromosome rupture can be induced with ethyleneimine; glyoxal, added during the S-phase, reinforces this action of ethyleneimine. This again suggests that glyoxal alters DNA during the replication process.

The question remains whether a brake on biological growth may actually exist in the form of a universally present and chemically active grouping, and whether it is possible to isolate biogenic α-ketoaldehydes possessing 'retine' activity and to identify them chemically.

Együd and Szent-Györgyi obtained from the liver of young calves several aldehydes which were bound to sulphydryl groups of proteins. These were liberated by treatment with arsenite and could be isolated with 2,4-dinitrophenylhydrazine as osazones[62,66]. 2-Keto-3-deoxyglucose was identified as one of the main components of this aldoketone fraction, and its constitution was confirmed by synthesis. This substance, however, is biologically inactive and thus does not correspond to the expected 'retine'. Of three further components, termed A, B, and C, A is the dominant component and inhibits protein biosynthesis even in very small amounts. This would be expected of an α-ketoaldehyde and could thus be the desired 'retine'. To date, its chemical constitution is unknown. The components B and C, present in small amounts, are biologically inactive.

Although these results appeared at first to be very interesting, they did not solve the problem of the existence of 'retine'-active biogenic α-ketoaldehydes. Otsuka and Együd[67] reported in 1968 that 2-keto-3-deoxyglucose is obtained from glucose spontaneously in the presence of arsenite, phosphate, and several amino acids, and that this reaction is not influenced by liver enzyme extracts. Four months later, and quite independently, Jellum[68] reported that a number of simple pentoses and hexoses gave, in the presence of arsenite and a phosphate buffer, of pH 7·6 at 37°C, several dicarbonyl compounds which, like ketoaldehydes, formed bis-(2,4-dinitrophenyl)hydrazones.

Since glucose, free amino acids, and phosphate are certainly present in adequate amounts in excised liver, it follows from the experiments of Otsuka and Együd and also those of Jellum that either ketoaldehydes isolated from liver after arsenite treatment *are* artefacts[67], or could be artefacts[68].

Recently Szent-Györgyi gave a report on the present state of his experiments in a lecture presented at Woods Hole[(2)]. It appears that Együd is still engaged in the isolation and identification of natural 'retine'. From the appropriate preparations methylglyoxal is always obtained as a hydrolysis product; it therefore still appears certain that 'retine' is a derivative of an aldoketone. Meanwhile further experiments are being undertaken with methylglyoxal, and there is new evidence that this compound is carcinostatic on local application[59,69].

An indication against the therapeutic use of systemic doses of methylglyoxal is the very active glyoxalase system of the blood. In order to protect methylglyoxal from the action of this enzyme system, nitrogen-containing compounds, such as simple amines, are brought into reaction with the aldehyde function of the methylglyoxal. The adducts thus obtained are virtually nontoxic and are able to develop their powerful carcinostatic effect even when given orally.

[(2)] We are very grateful to Professor Szent-Györgyi for sending us the manuscript of his lecture.

Quantitative methods for the determination of the aldoketone concentration in man are being further studied in the hope of developing a rationale for the administration of such carbonyl compounds in the treatment of malignant diseases.

In 1969 Kenny and Starkes[70] reported that certain human cells in culture produce a factor that selectively inhibits the growth of the less virulent strains of pathogenic bacteria. The bacterial virulence can then be interpreted as a resistance of the organism to this biogenic inhibitor. With reference to the 'retine' and 'promine' theory of Szent-Györgyi[62], in a subsequent paper[71] the authors examined whether in this case the inhibitory principle was again a ketoaldehyde. They pointed out that no biogenic ketoaldehyde recognizable as 'retine' has yet been isolated.

The first experiments with HeLa cells containing the factor showed that the growth of strains of *Staphylococcus aureus* of low virulence is inhibited. Addition of cysteine (1 mM) abolishes the inhibition, and although the inhibition of growth produced by the factor persists, that produced by methylglyoxal in concentrations below 1 mM disappears after a few hours.

From these initial results it does not as yet follow with certainty that the inhibitory factor is a ketoaldehyde. It does follow, however, that the factor acts through a reaction with thiol compounds which are of functional importance for growth, and that its inhibitory power could be stronger than that of methylglyoxal.

After separation from accompanying substances of high and low molecular weight, a freeze-dried preparation containing the inhibitory factor was obtained. Orange crystals were obtained from this extract, with 2,4-dinitrophenylhydrazine, taking care to avoid aerobic oxidation, and these crystals were identified as an osazone, indicating the presence of an hydroxyl function. It was thus concluded that an hydroxyketoaldehyde was present. The inhibitor was believed to be 4-hydroxy-2-oxobutanal, since this substance, when heated with acid, undergoes β-elimination of water to give vinylglyoxal, whose osazone showed spectral characteristics identical to those obtained from the above inhibitor extract after heating in acid solution. A direct experimental proof of identity would, however, still be required.

With this objective, model 4-hydroxy-2-oxobutanal was synthesized. Its 2,4-dinitrophenylhydrazone gave the same IR light absorption as that obtained from the inhibitor extract. The spectra of the 2,4-dinitrophenylhydrazones of the dehydrated model substance, and of the dehydrated inhibitor extract, were also identical. Investigations on the biological activity of the model substance have also been published, although it had already been concluded that the inhibitor present in HeLa cell culture medium was, in all probability, 4-hydroxy-2-oxobutanal. In a later paper[72] the authors reported that the model substance possesses the same biological properties towards selected strains of *Staphylococcus aureus* as does the naturally occurring preparation. The above conclusions were thereby confirmed, namely that human tissue culture cells produce 4-hydroxy-2-oxobutanal as a bacteria-inhibiting factor.

In summary, it appears certain that one or more factors present in HeLa cell-culture media inhibit the growth of the more weakly virulent pathogenic bacteria. One cannot yet say with certainty which substances are responsible for the inhibitory effects, since Kenny and Sparkes had added arsenite in their attempts to isolate the inhibitory factor, and since, on the basis of the results obtained by Otsuka and Együd[67] and by Jellum[68], the hydroxyketoaldehyde isolated by Kenny and Sparkes, may yet turn out to be an artefact.

5.1.5.3 *Miscellaneous*

Glyoxal given over a long period causes damage in mammalian organisms to the β-cells of pancreas islet tissue; this results in a raised insulin secretion and consequently in hypoglycaemia. Glyoxal also causes degenerative changes in the kidney[73-75].

Ling[76] examined the stabilizing effect of formaldehyde, methylglyoxal, glyoxal, and glutaraldehyde on suspensions of erythrocytes, which are required for the haemagglutination test. The most stable suspensions are obtained with methylglyoxal; the stabilizing action of glyoxal on haemoglobin has already been referred to in section 5.1.3. Procaccini et al.[77] reported that methylglyoxal and phenylglyoxal inhibit *in vitro* the acetylation of histones by a cell-free preparation from rat uterus, probably by deactivation of coenzyme A. This observation is of importance because reversible histone acetylation is a possible mechanism of the control of cell growth[78].

Herbage et al.[79] and Pallov et al.[80] found that methylglyoxal strongly inhibits mitosis in *Allium sativum* during prophase. The authors supposed that methylglyoxal in this case disturbs protein synthesis at the ribosomal level, and that certain functions of nuclear histones which are necessary for cell division are modified.

The sensitivity of bacteria and mammalian cells towards gamma irradiation can be significantly increased by α-ketoaldehydes[81-83]. The most active aldehydes are methylglyoxal, phenylglyoxal, and propyl-glyoxal; at concentrations of approximately $0.5-1.0$ mM they raise this sensitivity of bacteria (*Serratia marcescens*) by two to fourfold. These aldehydes act most effectively if added to the cell cultures before irradiation. The authors consider the sensitization to be based on a reaction of the aldehydes with sulphydryl groups in the cell.

5.1.6 Effects of α-ketoaldehydes on tumour growth

The results of Underwood and Weed[84] on the antiviral activity of α-ketoaldehydes first stimulated investigations on the antitumour activity of these compounds by French and Freedlander[85]. These investigations then gained a further sharp impetus when Együd and Szent-Györgyi[53] put forward the theory that α-ketoaldehydes are involved in the regulation of cell growth. The actions of glyoxal, methylglyoxal, and kethoxal on experimental tumours in animals were particularly investigated.

French and Freedlander[85] reported that kethoxal prolonged the survival of mice with leukaemia L 1210 by about 50%. Phenylglyoxal, cyclohexylglyoxal, and glyoxal are less effective. The action of methylglyoxal towards different experimental tumours of animals was described by Apple and Greenberg[69], Együd and Szent-Györgyi[86], and Jerzykowski et al.[87]. The results showed that methylglyoxal has general antitumour activity, that is, it is not limited to one tumour type. A single intraperitoneal injection of 80–125 mg/kg body weight suffices to increase sharply the survival time. Methylglyoxal inhibits the growth in mice of Ehrlich ascites carcinoma, carcinoma of the breast, adenocarcinoma, lymphosarcoma, and sarcoma-180. In hamsters complete regression of a Kirkmann–Robbins hepatoma takes place[87] if methylglyoxal is injected in the region of the tumour. A single injection of methylglyoxal produces in leukaemia L 1210 a 95%–99% reduction in the cell count. However, this is insufficient to achieve an appreciable increase in survival time on therapy, since the few surviving tumour cells are very virulent and divide about twice a day. This is a typical example of those substances which cause a drastic reduction of tumour cells in a single dose, yet do not bring about a significant prolongation of survival. Skipper et al.[88] has reported on the relation between survival time and the number of implanted tumour cells. In leukaemia L 1210, 10^9 cells/mouse are lethal, and it may be assumed that this order of magnitude may also be valid for other intraperitoneally growing tumours.

For the hypothetical case of a single implanted or surviving cell being 100% lethal, a linear relationship exists between survival time and the division cycle of the cell. Figure 5.5 shows that a tumour cell reaches the lethal number of 10^9 cells/mouse within fifteen days if the doubling time is twelve hours. For a tumour with twenty-four hour doubling time a cell divides to produce a lethal number only after thirty days. This shows clearly how difficult it is to treat rapidly growing tumours successfully by chemotherapeutic methods. It should certainly be possible to kill off the cells which survive an initial treatment by further doses of the antitumour agent, but in practice it has been shown, for example, with methylglyoxal, that daily treatment is no more effective than is a single dose. There are two possible explanations for this result. It may be that the surviving tumour cells are naturally resistant to the drug, and that this resistance is transmitted also to the daughter cells. Alternatively, it may be that the drug stimulates in the tumour cell or in the tumour-bearing animal the formation of adaptive enzymes, which then deactivate the inhibitory substance.

The mechanism by which α-ketoaldehydes inhibit tumour growth is not known in any detail. Since α-ketoaldehydes generally inhibit the growth of a wide range of cells, including healthy mammalian cells and bacteria, the tumour-inhibiting effect may be regarded as an expression of the general cytostatic property of these aldehydes. What has been said about the growth-inhibiting action in section 5.1.5.2 will be substantially valid here.

Although no exact comparative values are available, it may be assumed that tumour cells are more sensitive than healthy cells, although it has not been proven whether this is connected with a raised glyoxalase activity in tumour cells, as suggested by the theory of Szent-Györgyi.

Various experiments have been carried out to raise the selectivity of ketoaldehydes towards tumour cells. An interesting suggestion was made by French and Freedlander[85] and by Knock[89]. For tumour therapy these authors recommend the hydrogen sulphite complexes of the ketoaldehyde, and aldehyde hydrogen sulphite amine complexes (general formula: $RCOCHO \cdot NaHSO_3 \cdot NH_2R'$, where $R'NH_2$ is aminophenol or phenylenediamine). These complexes have the advantage of being less harmful than the free aldehydes. The significant advantage, however, is that with these complexes the action may be prolonged for a considerable period. The hydrogen sulphite complex is dissociable under physiological conditions, and the concentration of free aldehyde is lower by a factor of one hundred[90]. As the ketoaldehyde reacts *in vivo* with the components of the tumour, it is replenished by a further dissociation of the complex. Another advantage is that such polar anionic complexes tend to concentrate in the ground substance between tumour and tissue[91]. According to Knock[89] these ketoaldehyde complexes show a significant antitumour activity towards human cancers *in vitro*, for example, towards adenocarcinoma of the breast and the colon, and carcinoma of the epidermis.

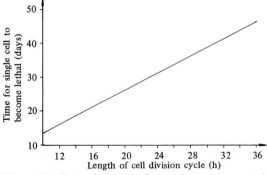

Figure 5.5. Rate of cancer cell growth versus expected survival time. Fast-growing cancers will kill in a shorter time than slow-growing ones if the concept of a lethal cell number[30] is valid. This plot shows the number of days a single cancer cell will take to reach the lethal number of 10^9 cells/mouse as a function of the length of the cell division cycle (according to Apple and Greenberg[69]).

A different, and very interesting, concept was put forward by Vince and Wadd[92], Vince and Daluge[93], and Jellum and Elgjo[94]. These authors supposed that inhibition of the metabolism of methylglyoxal in the cell would lead to an accumulation of methylglyoxal (or of other ketoaldehydes), which would then achieve an inhibition of cell growth. The growth-inhibiting effect of α-ketoaldehydes could be enhanced in this manner. In fact, the high lactate concentration (Issekutz[95]) and the deficiency in methylglyoxal (Lewis et al.[96]) indicate that tumour cells have lost their capacity to maintain the correct methylglyoxal balance.

α-Ketoaldehydes may be matabolized in the living cells by two enzyme systems, namely the glyoxalase system, and α-ketoaldehyde dehydrogenase. Active tumour inhibitors which depend on the principle of an intracellular accumulation of ketoaldehyde should in consequence have the capacity to inhibit both enzymes. The glyoxalase system is present in the cells of all life forms[97]. The natural substrate appears to be cytotoxic methylglyoxal, which in the presence of glutathione is converted into the nontoxic lactate (cf scheme 5.5). Vince et al.[92,93] have shown that S-alkyl- and S-arylglutathione derivatives inhibit glyoxalase I, which catalyses the reaction

$$\text{methylglyoxal} + \text{glutathione} \rightleftharpoons S\text{-lactoylglutathione},$$

thus leading to an accumulation of methylglyoxal. The inhibition is competitive with glutathione. The inhibitory activity increases as the chain length of the alkyl residue increases from methyl to octyl—which shows that hydrophobic regions of the enzyme have considerable importance in the binding of the inhibitor to glyoxalase I. The most potent glyoxalase inhibitors are S-arylglutathiones, which also have the advantage of being able to penetrate into the cell more readily than the alkyl derivatives. S-Arylglutathiones enhance the effectiveness of methylglyoxal against L 1210 leukaemia cells by a factor of fourteen to fifteen, as shown by determination of cell numbers in vitro. The investigations of Jellum and Elgjo[94] have shown that S-hexylglutathione strongly inhibits glyoxalase I but has no effect on the α-ketoaldehyde dehydrogenase, which converts ketoaldehydes directly into α-keto acids with the help of $NADP^+$ or NAD^+.

α-Ketoaldehyde dehydrogenase appears to be fairly nonspecific, since not only glyoxal but also hydroxymethylglyoxal and 2-oxo-3-deoxyglucose are converted into the corresponding α-keto acids[98,99]. The enzyme has been detected in the kidney and lung of sheep[98], in the liver of rat and mouse, and also in human adenocarcinoma[94]. According to Monder[98], rat kidney and lung do not contain the enzyme. It may be supposed that the tumour-inhibiting effect of α-ketoaldehydes can be enhanced by glyoxalase inhibitors only in those cells which possess little or no α-ketoaldehyde dehydrogenase activity.

5.1.7 Action of α-ketoaldehydes on the growth of bacteria

Schmidt[100] reported as early as 1931 that glyoxal could deactivate the toxin of diphtheria bacilli. It was later shown by Bloch et al.[101] that different 1,2-dicarbonyl compounds may exert a bacteriostatic effect on tuberculosis bacteria. Glyoxal and diacetyl proved to be the most efficacious, concentrations as low as 1 mM producing total inhibition of bacterial growth. Earlier authors[102,103] had assumed that diacetyl was the active principle of Finnish wood tar, used against tuberculosis in folk medicine.

Several carbonyl compounds occurring in milk, in particular glyoxal and diacetyl, inhibit test cultures of milk streptococci and other pathogenic bacteria[104]. The growth of *Staphylococcus aureus* and its lysis by Twort bacteriophages has been employed by German et al.[105] for testing antiviral substances. They reported that glyoxal inhibits bacterial growth, whereas the homologue glutaraldehyde has no influence on bacteria, but, on the other hand, stops bacteriophage activity. Good[106] recommends addition of glyoxal and glutaraldehyde to disinfectants in order to increase their antimicrobial activity.

The susceptibility of bacteria towards irradiation may be enhanced by a factor of two to four, if the cells are first treated with α-ketoaldehydes[81-83]. The growth of *Salmonella typhimurium* LT2 is inhibited by irradiated sugar solutions, glyoxal, methylglyoxal, and several α,β-unsaturated carbonyl compounds. This is discussed in detail in section 5.4. The inhibition of bacterial growth caused by α-ketoaldehydes probably follows the mechanisms referred to in section 5.1.5.2.

The inhibitory effect of methylglyoxal on the growth of *Escherichia coli*, already mentioned above, is also of special interest because it has been shown that methylglyoxal is a normal metabolite of *E. coli*[107,108]. *E. coli* contains an enzyme system, not as yet precisely characterized, which converts dihydroxyacetone phosphate into methylglyoxal. The methylglyoxal is then converted to D-lactate by glyoxalase, and then further to pyruvate by D-lactate oxidase. Through this reaction sequence *E. coli* can degrade glucose and glycerol to pyruvate via methylglyoxal. Normally this pathway is so exactly regulated by means of two control mechanisms that endogenous methylglyoxal cannot accumulate in cytotoxically active amounts. Freedberg et al.[108] have shown, however, that there are certain strains of *E. coli* which have lost this control mechanism, and which can produce increasing amounts of the bactericidal methylglyoxal from glycerol, thus killing themselves. In these mutants the particular enzyme that converts dihydroxyacetone phosphate into methylglyoxal also shows a very high activity. Mutants of *E. coli* have also been found whose resistance towards exogenous methylglyoxal is accompanied by a sharp increase in glyoxalase activity.

Barret et al.[109] have prepared over two hundred and thirty bisthiosemicarbazones of 1,2-dicarbonyl compounds, and have tested their biological activity. They found that the thiosemicarbazone of glyoxal is the most effective bacteriostatic derivative. Concentrations as low as $0.01-1.0$ μg/ml are effective in vitro against Trichomonas vaginalis or Trichomonas foetus.

5.1.8 Action of α-ketoaldehydes on viruses

The antiviral properties of α-ketoaldehydes and related compounds have been intensively and systematically examined in the research division of the Upjohn Company (Michigan), and the results published in a series of papers[84,110-114]. The first investigation, by Underwood, dealt with the question of whether α-ketoaldehydes are suitable for the sterilization of human blood and plasma. The danger of infection with viral diseases, serum hepatitis in particular, is a well-known major problem in transfusions. UV-irradiation, or treatment with β-propionolactone, or a combination of both methods[115], does not result in a satisfactory deactivation of the virus. Underwood's results, which are summarized in table 5.1, show that kethoxal and glyoxal are very effective virucidal agents and, in concentrations of 6.7 mM, are able to deactivate completely influenza virus, mouse hepatitis virus, and Newcastle disease virus (not tabulated). Since glyoxal does not lyse erythrocytes but, in contrast, as Suzuki and Hachimori[24] have shown, acts in a stabilizing manner and does not alter the oxygen uptake, this compound appears to be a suitable agent for the sterilization of blood.

Systematic investigation of one hundred or more different α-ketoaldehydes and structurally related compounds by the above mentioned research group, as well as by other authors[116-118], showed that in general five classes of compounds possessed antiviral activity:

(a) α-ketoaldehydes (for example, glyoxal, methylglyoxal, higher alkylglyoxals; kethoxal; phenylglyoxal, and other arylglyoxals);

Table 5.1. Comparison of the virucidal activities of different compounds towards influenza virus A (PR 8) and mouse hepatitis virus. The viruses were incubated for 2.5 h with 6.75 mM inhibitor in chick blood. The residual infectious units were determined by titration in eggs (influenza) or in mice (hepatitis) (according to Underwood[84]).

Compound	Lg titre	
	influenza virus	mouse hepatitis virus
Control	8.1	6.8
Methylglyoxal	7.4	—
Glyoxal	2.0	2.0
Kethoxal	2.0	2.1
Formaldehyde	5.3	—
β-Propionolactone	6.0	3.4

(b) vicinal di- and tri-ketones (for example, ninhydrin, diacetyl, cyclohexanetrione, 2,3,4-trioxopentane);
(c) α-hydroxyaldehydes (glycolaldehyde, glyceraldehyde, derivatives of β-aminolactaldehyde; lactaldehyde is inactive);
(d) α-ketoalcohols with primary hydroxyl groups (hydroxyacetone, 1,3-dihydroxyacetone);
(e) enediols (ascorbic acid, dihydroxyfumaric acid).

For the tests of virucidal activity the virus was generally preincubated for two to three hours at 37°C with the compound to be tested. Without prior removal of the excess of inhibitor, the residual infectious units were determined by titration in the chorioallantoic membrane of incubated eggs. Influenza viruses (PR8 and A-USA-47) and Newcastle disease viruses were mostly used for the tests. α-Ketoaldehydes are, however, also active against mouse hepatitis, polio viruses[114], MHV_3 viruses[117], and tobacco mosaic virus[34].

The activity of α-hydroxyaldehydes and α-ketoalcohols is remarkable. It may be supposed that these compounds are first converted into α-ketoaldehydes by dehydrogenating enzymes in the host cells. The enediols probably become active only after oxidation to vicinal dicarbonyl compounds. If the active groups are converted into stable derivatives (for example, the aldehyde group to an acetal, or a hydroxy group to an ether or an ester), activity is reduced. Dissociable derivatives, such as hydrogen sulphite adducts, retain their virucidal activity at least partially. It may therefore be concluded with some confidence that the α-dicarbonyl ·(C:O)·(C:O)· system is the vehicle of the virucidal activity. The antiviral properties of α-ketoaldehydes having been established *in vitro*, investigations were carried out on their curative effect in virus infected animals. Experiments with kethoxal resulted in no significant success with Newcastle disease virus in chicks, or influenza virus in mice. This is not difficult to understand since all α-ketoaldehydes react readily with cysteine, arginine, and also with other compounds present in the blood and tissues. The reaction products thus formed are largely inactive, as Underwood *et al.*[113] have shown. External treatment involving direct contact between the inhibitor and virus appears, however, to be successful. Thus Remis[119] could observe an appreciable reduction in para-influenza-3 virus on intranasal treatment with kethoxal.

Only a few experiments have been conducted to elucidate the mechanism of virus deactivation by α-ketoaldehydes. Sproessig and Muecke[120,121] ascertained that the concentrations necessary for virus deactivation are not toxic for the cells of the allantoic membrane. It can thus be ruled out that the reduction of the virus titre may only be a trivial consequence of a toxic effect on the host cell. An observation of Underwood *et al.*[113] also confirms that such an effect does not operate; the inhibitor is completely used up in the two to three hour preincubation period.

The most probable explanation is that α-ketoaldehydes react directly with arginine residues of the virus protein coat and with guanine bases of virus nucleic acids, thereby blocking or modifying the genetic information, so that a synthesis of intact virus in the cell is no longer possible (cf section 5.1.4, and Staehelin[34]).

5.2 Malonaldehyde
5.2.1 General features
Malonaldehyde (MA), when heated with 2-thiobarbituric acid reagent, forms a red–pink colour having a characteristic light absorption with a maximum at 530 nm[122-124]. This colour reaction is not very specific, since in oxidized fats, for example, other substances which also give a positive reaction with thiobarbituric acid may be present in small concentrations[125-127]. Nevertheless the method is most commonly used for the detection and quantitative determination of MA in foods or tissues, even though there is a very specific fluorimetric method for MA, based on its reaction with aromatic amines[128].

Malonaldehyde is one of the main carbonyl compounds that are formed by chemical or radiation-induced autoxidation of unsaturated lipids. It has been shown, by means of thiobarbituric acid, to be present in oxidized foodstuffs[122,123,129], methyl linolenate[125], methyl arachidonate and squalene[130], and in oxidized fatty acids[131]. It has also been shown that MA is formed on γ-irradiation of aqueous solutions of glycine[132], methionine, homocysteine, aspartic acid, and glutamic acid[133], and also by the action of ionizing radiation on glucose solutions[134,135]. γ-Irradiation of milk[136], meat[129,137], and other foodstuffs[130,134] causes formation of MA, so that its presence is often taken as proof that irradiation of the foodstuffs has taken place.

Numerous biological structures, such as cell membranes, mitochondrial membranes, and microsomes, contain highly unsaturated lipids which are accessible to oxidative changes, especially in the presence of oxidation catalysts such as haemoproteins. There are numerous indications that the free radicals, peroxides, and reactive aldehydes, which are formed by these changes, may be responsible for the ageing and destruction of cells and cell particles[138-141]. Since the autoxidation of lipids *in vitro* always produces MA, it may be supposed that this aldehyde is also formed *in vivo* in the autoxidation of cell lipids.

Malonaldehyde is very active and undergoes reactions with amino acids[143,152,155,157], proteins[142,152,164,170-176,180], and nucleic acids[144,149], but it can also be metabolized[145]. It is therefore not surprising that it has been possible to establish the formation of MA *in vivo* in only a few cases. MA is formed in erythrocytes[147] and in liver[148] if lipid peroxidation is induced by hydrogen peroxide or by carbon tetrachloride. When rats were given ethanol, the content of MA in liver was raised; it was concluded that MA is one of the causes of alcohol-induced liver damage[149,150].

Malonaldehyde (I), like acetylacetone and other 1,3-dicarbonyl compounds, has the tendency to form tautomeric enols, so that in aqueous solution MA exists as β-hydroxyacrolein[160-162]. As the enolic hydroxyl group has a pK value of 4·65, β-hydroxyacrolein exists in two different forms according to the pH. Below pH 4·65 the undissociated cyclic form (II) is largely present, stabilized by an intramolecular hydrogen bond of the chelate type[161]. Above pH 4·65 the proportion of the enolate anion (III) increases with increasing pH, and at neutral pH it becomes the main component (cf scheme 5.6). Stable salts may be prepared from the enolate form of MA. For biological experiments the sodium salt of

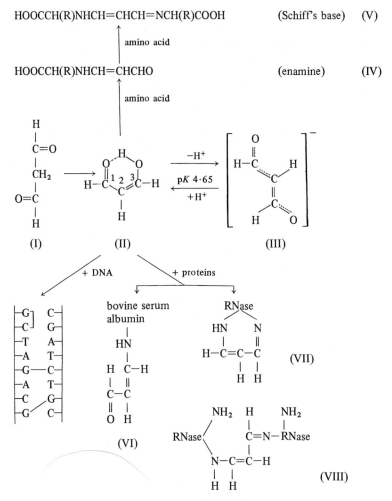

Scheme 5.6. Structure and reactions of malonaldehyde.

β-hydroxyacrolein is usually employed; it is obtained from 1,1,3,3-tetraethoxypropane[167]. Whilst the cyclic chelate form has the *cis*-configuration, the enolate anion has a *trans*-configuration with the negative charge being distributed over the whole molecule, as seen from NMR and IR investigations of aqueous solutions of MA[163].

The cyclic chelate form has a UV absorption maximum at 245 nm, $\epsilon = 13000$, and the enolate anion a maximum at 267 nm, $\epsilon = 30000$ [164]. The effect of the pH value on the yield of MA obtained on distillation shows that only the cyclic chelate form is volatile[162,165]. The behaviour of MA on Sephadex G10 columns is also of interest[166]. The cyclic chelate form of MA is very readily adsorbed on the gel matrix, because of its partial aromatic character. Aromatic and heterocyclic compounds are retarded by Sephadex more strongly than other water-soluble substances. The distribution constant (K_d) of MA alters continuously from 0·76 to 1·17 when the pH is lowered from 7 to 3. The reactivity of MA is strongly influenced by the pH value. The β-hydroxyacrolein present in the acid region reacts readily with nucleophilic groups (amino groups, sulphydryl groups) in 1,4-addition reactions, whereas the enolate anion chiefly present at neutral pH is appreciably less reactive towards nucleophilic reagents[143]. At neutral pH it is probably the undissociated form present in the equilibrium which undergoes reaction.

5.2.2 Reactions of malonaldehyde with amino acids

The mechanism of the reaction of MA with glycine was examined in 1966 by Crawford *et al.*[143]. The nucleophilic nitrogen atom reacts in a 1,4-addition reaction at the C-3 position of the α,β-unsaturated carbonyl system. Water is eliminated from the unstable β-hydroxyaldehyde, and the enamine [(IV) in scheme 5.6] is formed. Leucine, valine, the ethyl esters of leucine and valine, and hexylamine react in basically the same manner as glycine to give the corresponding enamines[152].

Enamines are themselves reactive substances and may react with a second molecule of amino acid, amino acid ester, or amine to give a Schiff's base of the type, N,N'-disubstituted 1-amino-3-iminopropane [(V) in scheme 5.6][152].

The preparation of the enamines and Schiff's bases took place in hydrochloric acid solutions, but it may be inferred that these same reactions also occur at neutral pH, since both the 1,4-addition reaction and the formation of Schiff's bases can be catalysed by hydroxyl ions. Aqueous solutions of N,N'-disubstituted 1-amino-3-iminopropane derivatives [formula (V)] are yellow, possessing a characteristic light absorption with well-defined maxima at 256 and 435 nm, and diffuse maxima at 280 and 370 nm [152]. On account of the 6 π-electron chromophoric system, N=C−C=N, an intense blue fluorescence is observed with an excitation maximum at 370 nm and an emission maximum at 450 nm [128,152].

The age pigments, ceroid and lipofuscin, found in the cells of heart muscle and the liver, are generally regarded as reaction products of lipid oxidation[153]. From these pigments a fluorescent component could be isolated, whose chemical, IR spectrometric, and chromatographic properties indicated that it was an autoxidized cephalin[154,155]. It may be supposed that on autoxidation of the polyunsaturated fatty acid components of membrane lipids high concentrations of MA may be formed locally. This MA is not metabolized but instead reacts, at the site of formation, with primary amino groups of the cephalin bases to give yellow compounds. These may show the same fluorescence properties as the Schiff's bases derived from MA and amino acids[155-157].

Buttkus[266,267] has studied the reaction of cysteine and methionine with malonaldehyde. Methionine reacted at the amino group, and the elemental analysis was in agreement with the structure of an enamine [formula (V) in scheme 5.6]. From elementary analysis and titration studies it was concluded that 2 moles of cysteine react with 3 moles of malonaldehyde to give a product with the formula

$$O=CH-CH_2-CH(S-CH_2-CH-COOH)_2$$
$$|$$
$$NH-CH=CH-CH=O$$

It may be supposed that all sulphydryl compounds, which are more nucleophilic than amino groups[158], likewise react with the enol form of MA (β-hydroxyacrolein) in a 1,4-addition reaction. This view is supported by the fact that acrolein and all other 2-alkenals react very rapidly with sulphydryl compounds (section 3).

5.2.3 Reactions of malonaldehyde with nucleic acids

Experiments with ^{14}C-labelled MA[144], obtained by irradiation of [^{14}C] glucose solutions, have shown that MA reacts directly with the amino (NH_2) groups of the guanine and cytosine bases of purified DNA from calf thymus. The mechanism of the reaction presumably corresponds to the reactions with amino acids and aromatic amines. With these bases cross-linkings take place within a single strand of DNA and/or between two DNA strands of a double helix (scheme 5.6). As is known from investigations with other bifunctional agents[159], this leads to a profound alteration in the structure of the DNA. Thus, after MA treatment, DNA shows a reduced hypochromia, a profound change in the temperature-absorption curve, and an increased resistance towards degradation by DNase. Similar alterations are also shown by DNA from human fibroblasts which have been incubated with MA (50 μg/ml) for five hours.

Incubation of rat liver DNA with MA *in vitro* (molar ratio 100:1) results in a complete loss of template activity. An effect of MA on DNA could also be established for experiments *in vivo*[168]. Daily doses of 100 μg of MA/g body weight administered to rats orally or intraperitoneally similarly produce a decrease in the template activity of the liver DNA.

This decrease depends on the age and weight of the animals, and the effect is greatest for young animals and at the beginning of the MA treatment. This suggests that older animals, or animals treated with MA for a longer period, develop enzymes which can metabolize and detoxify MA, so that DNA damage becomes less. These MA-metabolizing enzymes are presumably aldehyde dehydrogenases and aldehyde oxidases of low specificity, which cause low concentrations of MA to be oxidatively degraded by rat liver mitochondria[146].

5.2.4 Reactions of malonaldehyde with proteins and enzymes

Kwon and Brown[142] have shown that MA forms a stable complex with bovine serum albumin (BSA). It was assumed that a BSA molecule possesses sixteen to seventeen specific binding sites for MA, and that in this reaction it is the free amino groups of the protein which undergo reaction. A closer examination of the reaction showed[170] that MA reacts with BSA with first-order reaction kinetics and that the reaction is strongly dependent on the hydrogen ion concentration, with the reaction velocity reaching its maximum at pH 4·3. The reaction probably proceeds according to the same mechanism as the reaction with glycine (scheme 5.6). The ϵ-amino group of lysine and the amino group of the N-terminal aspartic acid react with the undissociated enol of the MA. The enamine derivative [formula (VI), scheme 5.6] thus formed is the major reaction product. The quantitative relationship between the amount of bound MA and the loss of ϵ-amino groups of lysine and of N-terminal aspartic acid indicates that MA reacts with only one amino group. This also explains the surprising result that MA, in contrast to other dialdehydes (such as glutaraldehyde), gives no intermolecular cross-linking reactions with BSA[170]. Even when MA reacts with gelatin solutions, the latter do not increase in viscosity. This confirms that no cross-linkings are formed here, viscosity measurements being commonly used to demonstrate cross-linking reactions. It was also shown[164] that proteins in foodstuffs react with MA at pH 6·5–7·1, and that even enzymes like lipase[171] are deactivated at pH 7·0.

Treatment of wheat flour gliadin with MA leads to a modification of the total tyrosine, as shown by amino acid analysis on the protein after hydrolysis with 6 M hydrochloric acid[172]. Other authors found that MA treatment of gliadin leads to a loss of lysine and tyrosine[173], and of cysteine and methionine[174,175]. The discrepancy between these results is presumably explained by the different conditions used for the hydrolyses. In strongly acid media the enamines formed from primary amines and MA are split again into their components; this cleavage is strongly catalysed by increasing the hydrogen ion concentration. Loss of lysine and other amino acids can therefore be determined only if these amino acids have been specially labelled before the hydrolysis. For the labelling of amino groups, which remain free after MA treatment, 2,4-dinitrofluorobenzene has been employed. The reaction of MA with myosin has been examined[176] in connection with the possible importance of this reaction for protein denaturation in the low temperature storage of fish. Curiously, the ε-amino groups of myosin react with MA more rapidly at $-20°C$ than at $0°C$. The amino acid analysis of the myosin treated with MA below $0°C$ shows that lysine, tyrosine, methionine, and arginine have reacted in that decreasing sequence. The raised reactivity at $-20°C$ was explained by a catalytic effect of the ice structure.

The influence of MA on ribonuclease (RNase) has also been closely investigated[152,180]. When RNase (10 mg) was incubated in MA (4 mM at pH 7·0) a complete loss of activity occurred within 100 hours. Amino acid analysis showed a loss of one molecule of methionine, three molecules of tyrosine, and two molecules of lysine per molecule of RNase. The values for normal RNase are four molecules of methionine, six of tyrosine, and ten of lysine[178]. One lysine residue is essential for the structural integrity and for the enzyme activity[179], and the loss of activity after MA treatment shows that this particular lysine residue has reacted with MA, whereas the loss of methionine has no influence on the activity. That MA affects both inter- and intramolecular cross-linkages [formulae (VII), (VIII), in scheme 5.6] is shown in the behaviour of MA-treated RNase on gel filtration with Sephadex G100, and from the fluorescence properties of the reaction products. This is surprising in view of the fact that such cross-linkages were not observed with BSA or with gelatin solutions [cf scheme 5.6, (VI)].

The mechanism of the reaction is probably the same as for the reaction with glycine. The amino group of a lysine residue reacts first with a molecule of MA; the enamine derivative formed then reacts with a second amino group of the same (or a different) molecule of RNase. The Schiff's bases formed thereby possess the conjugated double bond system which is responsible for the characteristic fluorescence properties of the various monomeric, dimeric, and higher aggregates of the MA-treated RNase. The excitation maximum occurs at 395 nm and the emission maximum at 470 nm. These spectral properties correspond absolutely to those possessed by Schiff's bases obtained from MA and amino acids or from α-N-acetyl-lysine.

MA inhibits the activity *in vitro* of lipase[49]. It is quite possible that deactivation of lipases *in vivo* may also occur after absorption of rancid fats, which always contain MA.

The fact that MA may react with proteins and enzymes under physiological conditions is of great biochemical and physiological significance. Protein-containing foodstuffs may be materially altered and denatured by MA through such reactions, since MA may have formed as the result of lipid autoxidation or through irradiation. MA may also be formed at the cellular level by spontaneous autoxidation of unsaturated cell lipids, and it may then alter cellular proteins and enzymes. Membrane structures are rich in unsaturated lipids and particularly phospholipids;

therefore MA formed by lipid autoxidation may produce far-reaching denaturation of cell structures. The possible role of MA in the formation of age pigments, which have the same spectral properties as the reaction products between MA and RNase or amino acids, has already been mentioned above.

Harman[141] believes that the interaction of autoxidized lipids with proteins plays a significant role in the ageing of cells and organs. Although numerous highly reactive substances, including radicals, are formed during lipid oxidation, it may be assumed that the majority of the effects of autoxidized lipids are due to MA, which is nearly always one of the major active products.

It should be pointed out that the initial reaction products between MA and proteins, namely the enamine derivatives, are themselves reactive compounds and can be alkylated or arylated by several biologically occurring electrophilic substances[177].

5.2.5 Effects of malonaldehyde on various cell functions

Malonaldehyde stimulates oxygen uptake by rat liver homogenates[181] and by rat liver mitochondria, which metabolize the aldehyde[182]. The interaction of MA with rat liver mitochondria has been examined in detail by Horton and Packer[146,169]. These mitochondria contain aldehyde oxidases or dehydrogenases of low specificity which can oxidize MA, so that at low concentrations MA can serve as substrate for mitochondrial respiration.

The enzyme system has virtually the same affinity for MA as it does for acetaldehyde, as can be seen from the Michaelis constants (MA, $0\cdot 45$ mM; acetaldehyde, $0\cdot 23$ mM). As with hepatic aldehyde oxidase preparations[183], one finds differences between the mitochondrial oxidation of MA in males and females. The velocity of oxygen uptake by the mitochondria of female rats is much higher than that of the mitochondria of the male animals. With uncoupled mitochondria magnesium (Mg^{2+}) and manganese (Mn^{2+}) ions stimulate the oxidation of MA. As other ions (Na^+, K^+, Ca^{2+}, SO_4^{2-}) have no effect, one must regard the two ions (Mg^{2+}, Mn^{2+}) as specific cofactors. ATP and inorganic phosphate appear neither to be essential, nor to act as stimulators. At higher concentrations (4%–6%) MA reacts with mitochondrial proteins with the formation of cross-linkages and thereby stabilizes the mitochondrial membrane, as shown by the increased resistance towards detergents (Triton). However, the effect is not nearly as large as it is with glutaraldehyde, which not only stabilizes, but also preserves, isolated mitochondria. MA-stabilized mitochondria no longer show oxidative phosphorylation, although electron transport with succinate is not affected.

Malonaldehyde is also rapidly metabolized in the living rat[145]. Only a short time after intraperitoneal injection of 6 mg of MA/rat, MA can be detected in the serum and in the organs. However, whereas the aldehyde soon disappears from the serum, it is still detectable in the organs three hours later. The MA concentration is different in the various organs, and this fact appears to be connected with differences in its rate of oxidation.

It has already been mentioned that MA is not only metabolized *in vivo*, but also influences the liver DNA, as it does *in vitro*.

When rats are treated with ethanol[149,150] or carbon tetrachloride[148] there is a raised concentration of MA in the liver as a consequence of the resulting liver damage. This MA is formed in the lipid autoxidation reactions induced by ethanol or carbon tetrachloride.

The formation of MA was also observed in homogenates prepared from fatty liver[188]. It would appear that in this case the liver is no longer able to metabolize the aldehyde, which results in its accumulation. The aldehyde then alters the membrane proteins, enzymes, and also the nucleic acids, and is thus responsible, at least in part, for liver damage.

Quash and Taylor[185] have shown that β-aminopropionaldehyde is a component of human serum. In animal tissues this aldehyde is formed by enzymatic breakdown of spermidine, and it may be converted into MA by amine oxidases. These authors believe that MA is an essential factor for the control of cell growth, since it blocks the template function of DNA[168]. A decrease in β-aminopropionaldehyde oxidation leads to its own accumulation, to an accumulation of spermidine, and to a lowering of the MA content. A raising of β-aminopropionaldehyde and spermidine causes an increase in RNA synthesis[186,187], whereas a decrease in MA leads to an increase in DNA synthesis. It is possible that MA is the natural inhibitor of cell growth (retin), as postulated by Szent-Györgyi (cf section 5.1).

5.3 Glutaraldehyde
5.3.1 General features
Glutaraldehyde does not occur in living matter, and no evidence exists that it may be formed, like many other aldehydes, by spontaneous or induced degradation of the building blocks of cells. Nevertheless, glutaraldehyde is of great biochemical interest, as the large number of relevant papers demonstrates.

One of the most characteristic and most important properties of glutaraldehyde is the cross-linking effect on proteins. Because of its ability to produce cross-linkages in proteins, glutaraldehyde is much used as a fixative for electron microscopy and cytochemistry. Glutaraldehyde is also suitable for the stabilization of enzymes, cells, or cell particles, and for the coupling of proteins to other proteins or to cells. Many crystalline proteins can be stabilized and made water-insoluble with glutaraldehyde without the crystal structure being measurably altered. Protein crystallographers are therefore using glutaraldehyde-stabilized protein crystals increasingly for conformational studies. Because of its tanning action on collagen, glutaraldehyde is used on a large scale in leather manufacture. It also has excellent antimicrobial activity and therefore has been recommended for the chemical sterilization of hospital implements.

A knowledge of the structure of glutaraldehyde in aqueous solution is of considerable importance for the explanation of its cross-linking action. The molecular structure of glutaraldehyde in aqueous solution remains uncertain, although many studies have been published on this topic[189-192]. The tendency of glytaraldehyde towards polymerization, condensation, and hydration, complicates the situation considerably. The most relevant work appears to be that of Hardy et al.[192]. These authors came to the conclusion, on the basis of NMR spectroscopic investigations, that in aqueous solutions of glutaraldehyde an equilibrium situation obtains, as presented in scheme 5.7. The nonhydrated free aldehyde (I) is present in only very low concentrations in the equilibrium. The other three forms are distributed in the approximate proportions (II) 28%, (III) 40%, (IV) 32%, these values being independent of the absolute glutaraldehyde concentration. In the IR spectrum of glutaraldehyde in aqueous solution a C=O vibration is lacking, showing that the aldehyde groups are largely hydrated[193]. In neutral aqueous solution the equilibrium given in scheme 5.7 remains unaltered even on storage for several weeks. In weakly alkaline solution, however, polymerization and condensation take place very readily. Aldol condensations can lead to the formation of α,β-unsaturated aldehydes of types (V) and (VI), and higher polymeric products[191]. The existence of such products also follows from the UV light absorption. Pure glutaraldehyde solutions have a single maximum at

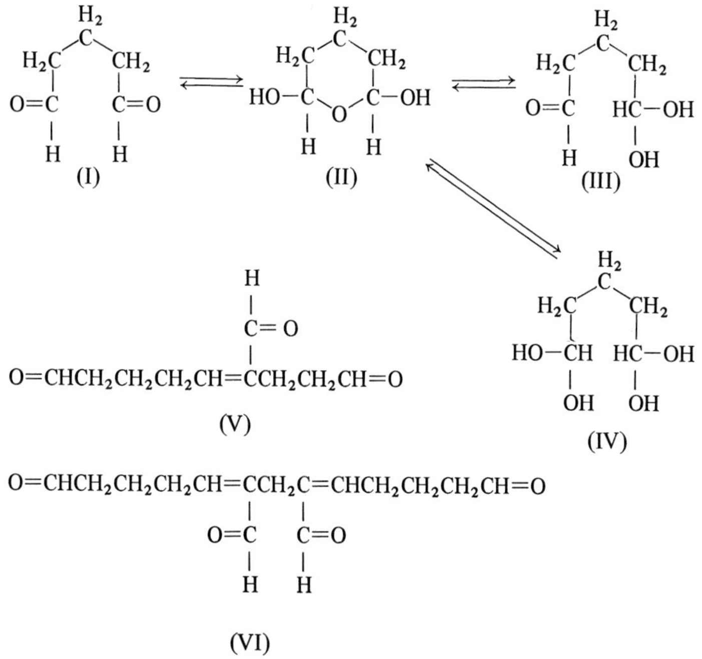

Scheme 5.7

280 nm, corresponding to the carbonyl absorption of forms (I) and (III). The intensity of this maximum increases with a rise in temperature and with time of storage[194,195]. If aldol condensation takes place, a second maximum develops at 235 nm, and the intensity of this maximum rises with increasing content of α,β-unsaturated aldehyde. A characteristic UV absorption maximum in the region 235-210 nm is given by all α,β-unsaturated aldehydes (cf section 3.2). Commercially available glutaraldehyde always contains a small proportion of α,β-unsaturated aldehydes, as the UV absorption at 235 nm indicates, as well as other polymerization products.

If glutaraldehyde is to be used for histochemical investigations of enzymes or for specific reactions with proteins, these impurities must first be removed because, as α,β-unsaturated carbonyl derivatives, they likewise react readily with proteins. It has been shown that the activity of different enzymes in glutaraldehyde-fixed muscle is inhibited the more strongly, the greater the amount of accompanying impurities with a light absorption maximum at 235 nm [194].

Vacuum distillation[194,196], treatment with active charcoal[194], or separation with Sephadex G10 [195,197] are all suitable for the purification of glutaraldehyde. Purified glutaraldehyde is stable for long periods if kept at 4°C at pH 5·0 under nitrogen and can be used for biochemical purposes[196,198].

5.3.2 Reactions with proteins

In spite of the great number of investigations on the reaction of glutaraldehyde with proteins, only very few results have been obtained on the reactions of this aldehyde with the lower molecular weight building blocks of proteins, namely amino acids and peptides. Some such experiments were carried out by Habeeb in 1968 [199], and Hopwood et al. in 1970 [200]. These investigations showed that glutaraldehyde in 0·5% solution at neutral pH reacts within thirty minutes with N-terminal amino groups of peptides (glycyltyrosine, leucylglycylphenylalanine, glycylglycylvaline, polylysine), and with the α-amino groups and sulphydryl groups of cysteine. The ε-amino groups of lysine similarly react quantitatively. In histidine, phenylalanine, and tyrosine, reaction occurs partly at the α-amino group and partly at the histidine or phenyl residue, and this is accompanied by an increase in UV absorption in the region 250 nm-300 nm. Alkaline glutaraldehyde solutions react with most amino acids to give a red colour the intensity of which is proportional to the aldehyde concentration[201]. The structures of the reaction products between glutaraldehyde and amino acids or peptides have not been investigated, but it is generally assumed that glutaraldehyde initially forms Schiff's bases with reactive amino groups. These Schiff's bases are then stabilized in a secondary reaction by a mechanism as yet unknown. A large number of papers has appeared on the cross-linking action of glutaraldehyde on collagen and gelatin[202-210], serum albumin[199,211,212], ovalbumin[199], and other proteins[211,213].

As shown for model compounds it is the free amino groups of proteins which react with glutaraldehyde, and it is the lysine residues which are the most important[202,214]. Thus, in BSA approximately 90% of the free amino groups react with glutaraldehyde[199]. For collagen the proportion is 82%[202], and for ovalbumin 50%[199]. Amino acid analysis of glutaraldehyde-treated carboxypeptidase A revealed that only the ε-amino groups of lysine had reacted, and of the fifteen lysine residues of the molecule ten had been modified[215].

Concentrated protein solutions (above 2%) give a yellow colour in their reaction with glutaraldehyde. Some proteins form a gel within seconds, some similarly give insoluble precipitates, and others give neither result even after twenty-four hours.

During the reaction with glutaraldehyde a characteristic alteration of the UV spectrum takes place. The absorption in the region 275 nm–300 nm increases strongly, and the protein maximum at 280 nm suffers a blue shift of approximately 5 nm. These spectral changes are believed to be due to a reaction of glutaraldehyde with the aromatic amino acids[199,200,216]. In salmine, which contains no tyrosine and no tryptophan, no such spectral alterations were observed with glutaraldehyde[199].

The reaction of glutaraldehyde with aromatic residues is reversible, and the compounds formed are not particularly stable, so that even on dilution glutaraldehyde is again formed by dissociation. The instability of these linkages is probably also the cause of the differences in results reported for the extent of the modification of the aromatic residues by means of glutaraldehyde. If amino acid analysis is carried out on acid-hydrolysed proteins, the content of aromatic amino acids determined is practically unchanged[215], whereas the direct determination shows a decrease[199].

The formation of gels and precipitates is connected with the cross-linking of protein molecules with glutaraldehyde. A survey of the cross-linking effect which bifunctional reagents have on proteins has appeared[217], and the significance of such reactions for histochemical examinations is discussed there. In the case of the bifunctional reagent, glutaraldehyde, the cross-linking depends mainly on the number of the lysine residues and aromatic amino acids present and their distribution at the surface of the protein molecule. If these groups are evenly distributed on the surface of the molecule, then intermolecular cross-linking is favoured; if these groups are built into the interior of the molecule, they will form mainly intramolecular cross-linkages. Such intramolecular linkages are also favoured by low protein concentrations and a high glutaraldehyde concentration. In a glutaraldehyde-treated protein, intermolecular and intramolecular cross-linkages will both be present, and their distribution will depend on the nature of the protein and the reaction conditions. In the case of glutaraldehyde-treated BSA, ovalbumin, and γ-G-glubulin, separation on Sephadex G200 has shown that only a small proportion of the monomer is present, whereas the main component consists of aggregates with perhaps three times the sedimentation constant[199]. Crystals of carboxypeptidase A, on treatment with glutaraldehyde, give an insoluble mesh of high mechanical stability. Its enzyme activity is not substantially altered, and the X-ray diffraction patterns of the treated and the untreated crystals are nearly identical[215,218]. Such insoluble, stable enzyme crystals are valuable for X-ray structural examination of enzyme–substrate complexes. A 0·05% glutaraldehyde solution partially deactivated glycogen phosphorylase b within ten minutes[219]. Its amino acid analysis showed that only 10% of the lysine residues present had reacted. The sulphydryl group had not reacted, as shown by determination with Ellman's reagent. With the aid of electrophoresis and Sephadex G200 a pure, stable enzyme which retained 65% of the activity of the normal enzyme could be obtained, and its affinity towards glucose 1-phosphate and glycogen remained unaltered. The modified enzyme is appreciably more thermostable than the normal enzyme.

Intermolecularly cross-linked complexes containing different proteins may also be prepared with glutaraldehyde. Such heteropolymers are probably the main products formed in the glutaraldehyde fixation of biological preparations for histochemistry and electron microscopy. It is probable that other reactive molecules (such as polysaccharides) also participate. Such heteropolymers are also of interest for the examination of antibody–antigen interaction. An ovalbumin–BSA conjugate gives a precipitate with anti-BSA and also with anti-ovalbumin, in other words the antigen-active groups of BSA and ovalbumin are found on the surface of the protein conjugate[199,221], and have not been altered by glutaraldehyde treatment.

Glutaraldehyde is also used to couple ACTH to BSA. This complex no longer shows any ACTH-like properties but, when introduced into rabbits, antibodies form against BSA and ACTH[222]. Various enzymes (peroxidase, glucose oxidase, tyrosinase, ribonuclease) have been bound to proteins with glutaraldehyde. Such conjugates possess the immunological activity of their components and also still possess a part of the original enzyme activity[223,224]. If glutaraldehyde is added to a solution of antibodies and antigens, insoluble complexes are obtained. These complexes are active, specific, and stable immunoadsorbents, which may be employed for the isolation of antigens and antibodies in column or batch processes[225]. Glutaraldehyde is also suitable for the binding of antibodies to erythrocyte membranes, this binding proceeding rapidly and irreversibly. Such antibody-laden erythrocytes are well-suited for the haemagglutination test, and very small concentrations of antigen may thus be determined[226-229]. The use of glutaraldehyde for the fixing of samples for electron microscopy and cytochemistry is not discussed further, since several excellent review articles exist on these topics[220,230,231].

As already mentioned, glutaraldehyde reacts with proteins chiefly at the ε-amino group of the lysine residues. The structure of the new linkages is not yet clear. The primary reaction is probably the formation of Schiff's bases; such linkages are not, however, particularly stable, and in any case are split by acid hydrolysis. It must therefore be supposed that the Schiff's bases are stabilized by secondary reactions, which may possibly take place only during the acid hydrolysis. From glutaraldehyde-treated collagen a compound could be isolated, after acid hydrolysis, which contains three glutaraldehyde residues per lysine residue[202]. A ratio greater than unity could also be found with other proteins for which the bound glutaraldehyde molecules per lysine residue could be determined. These binding ratios probably result from participation in the reactions of the dimeric or trimeric forms of glutaraldehyde [formulae (V) and (VI) in scheme 5.7].

5.3.3 Effects of glutaraldehyde on various cell functions

Isolated mitochondria[232-234] and chloroplasts[235-237] may be fixed with high concentrations of glutaraldehyde (250 mM) in such a manner that the ultrastructures and the conformation remain intact. Detergents like Triton X-100 produce no conformational changes in glutaraldehyde-fixed mitochondria, and electron transport and the capacity to produce ionic gradients are also not affected. However, various other energy-dependent processes are inhibited, for example, stimulation of respiration by dinitrophenol, calcium ions, and ADP, and the synthesis of ATP. The stabilizing effect on the configurational state of the mitochondria can still be detected in concentrations down to 5 mM. In still lower concentrations (from approximately 1 mM) glutaraldehyde is rapidly oxidized by liver mitochondria, and may serve as substrate for the respiration[238,239]. The rate of oxygen uptake is about equal to the oxidative metabolism of 1 mM succinate, and is two to three times greater than with other aldehydes (formaldehyde, acetaldehyde, glyoxal, MA). The glutaraldehyde oxidation is energy coupled, and is therefore stimulated by ADP and an uncoupler, and inhibited by oligomycin. The oxidation probably proceeds, as with other aldehydes, by means of unspecific aldehyde dehydrogenases. In human blood serum a formyl hydrate dehydrogenase has been detected, which in addition to formaldehyde, acetaldehyde, propionaldehyde, butyraldehyde, isobutyraldehyde, and crotonaldehyde, also oxidizes glutaraldehyde[241].

Glutaraldehyde fixation affects not only the membranes of mitochondria and chloroplasts; as has been reported above, glutaraldehyde binds antibodies to the erythrocyte membrane. Glutaraldehyde-treated erythrocytes are generally more stable than untreated erythrocytes, and may be stored unaltered for six months. On fixation the erythrocytes take up glutaraldehyde very rapidly in the first hour only; on further treatment no more glutaraldehyde is taken up even after several days.

In this way glutaraldehyde is bound not only to the amino groups of the cell membrane but also to the lysine residues of haemoglobin. Selective permeability of the cell membrane towards ions is lost on glutaraldehyde treatment[242].

In sarcoplasmic membranes calcium efflux is lowered after glutaraldehyde treatment[243]. The adenylcyclase system[244] and the ATPase, ADPase, and AMPase[245] are inhibited in plasma membranes (mitochondria, microsomes) of rat liver. The cell walls of several bacteria (*Bacillus subtilis*, *Bacillus licheniformis*) may be cross-linked with glutaraldehyde, they then become resistant towards degradation by lysozymes[246]. It was assumed that glutaraldehyde cross-links the free ε-amino groups of diaminopimelic acid, which are involved in the buildup of mucopeptides.

5.3.4 Action of glutaraldehyde on viruses and bacteria

In 1960 Ross[247] pointed out that glutaraldehyde solutions are very suitable for the chemical disinfection of surgical instruments. Since then numerous papers have appeared on the deactivating effect of glutaraldehyde on viruses, bacteria, bacteriophages, and fungi. In common with other aldehydes, glutaraldehyde exhibits biocidal activity only in experiments *in vitro*, and therefore is not used in the treatment of viral or bacterial illness. The antibacterial activity of glutaraldehyde is twenty to one hundred times greater than that of formaldehyde, according to bacterial type[248]. The antimicrobial activity of glutaraldehyde solutions is appreciably stronger in weakly alkaline media. 2% Glutaraldehyde solutions are used for the sterilization of equipment used in surgical operations and for hospital installations which are not sterilizable by physical methods[249-252]. Glutaraldehyde has the advantage over other chemical sterilizing agents in that it is not corrosive, does not give off volatile components, is of low toxicity, and acts within a short time. Thus, by means of a 2% glutaraldehyde solution at pH 7·5–8·5 non-spore-forming bacteria are killed inside two minutes, tuberculosis bacteria (*Mycobacterium tuberculosis*) within ten minutes, and spore-forming bacteria within three hours[200]. The deactivation of the following bacteria and viruses has been reported separately: polio viruses[253], vaccinia viruses[253,254], myxo viruses[255-257], yellow fever virus[256], influenza virus[259], Newcastle disease virus[258], vesicular stomatitis virus[260], bacteriophages[261,262], *Mycobacterium tuberculosis*[250], *Staphylococcus aureus*[248,261,263], *Escherichia coli*[264,265], and *Pseudomonas* species[264,248]. Very little is known about the mechanism of the virus deactivation, but it may be supposed that reactions with specific amino acids of the protein components play a decisive role. Glutaraldehyde-deactivated viruses have lost their infectivity, yet their immunological properties are unaltered[256,257]. This is understandable in view of what has been said in section 5.3.2. Considering that the preferred reaction of glutaraldehyde is with amino groups, it might well be expected that it would also react with nucleic acids in a corresponding manner, and thus fix or modify specific structures and alter biochemical and biological properties, just as MA is known to do (section 5.2). No experimental results relevant to these aspects appear to have been published yet.

5.4 α,β-Unsaturated carbonyl compounds from energetically irradiated aqueous sugar solutions

In their informative investigations on the biological effects of high-energy-irradiated foodstuffs, Schubert and Sanders[268] reported that aqueous sugar solutions which had been treated with ^{60}Co γ-rays strongly inhibited the growth of *Salmonella typhimurium* LT2. Since irradiation occurred in the absence of oxygen, these effects cannot be based on malonaldehyde, formaldehyde, glyoxal, or peroxides—the concentrations of these substances formed in unbuffered oxygen-free media are far too low, according to the data of Schubert and Sanders. As the irradiation dose is increased so the growth inhibiting action on bacteria of such sugar solutions increases. Simultaneously the carbonyl content of the solutions increases linearly, so that it would seem justifiable, however, to consider that the inhibitory effects are due to carbonyl compounds.

If the solution is heated under pressure in an autoclave (120°C, 15 lb in^{-2}, 20 min) *after* irradiation, the effectiveness increases. Addition of substances with free amino or sulphydryl groups reduces the cytotoxic activity, or abolishes it.

Since peroxides must be excluded for the reasons given above, the presence of carbonyl compounds, possibly also of compounds with conjugated double bonds, are again indicated. The UV absorption curves[269] confirm this. The next problem was the closer identification of these carbonyl compounds with antibacterial activity. The first attempt was by means of an empirical comparison with activities of substances of known chemical constitution. To achieve an exact basis of comparison a formula which had been developed earlier[270] and which presents the relation between inhibition of cell growth and the concentration of inhibitor as a linear function was used. The quantity referred to is that concentration of inhibitor (c_{10h}), which extends the doubling time, as measured without addition of inhibitor, by ten hours. For example, with glyceraldehyde as the model substance, without autoclave treatment a c_{10h} value of $23 \cdot 3 \times 10^{-8}$ μmol/cell was obtained; after autoclaving the value was $3 \cdot 6 \times 10^{-8}$ μmol/cell. Since methylglyoxal is formed by the dehydration of glyceraldehyde (see section 4.1), it seemed obvious to make UV spectrometric determinations of the content of methylglyoxal in these solutions and to calculate the c_{10h} values of glyceraldehyde on the basis of the methylglyoxal. This gave a value of $0 \cdot 9 \times 10^{-8}$ μmol/cell, which corresponds to the value for pure methylglyoxal. This confirms the supposition (section 4.1) that the source of the inhibitory activity of aqueous glyceraldehyde solutions is methylglyoxal.

Among their investigations of different model substances Schubert and Sanders[268] found that α,β-unsaturated carbonyl compounds of the general type RCH=CHCOR' possess c_{10h} values of $0\cdot2 \times 10^{-8}$–$1\cdot4 \times 10^{-8}$ µmol/cell, whereas a group of substances (as represented by glyoxal or acrylic acid) possess c_{10h} values which are distinctly separate, being in the region of 6×10^{-8} µmol/cell. Saturated or nonconjugated unsaturated compounds are much less active as inhibitors (figure 5.6). Some α,β-unsaturated carbonyl compounds are only weakly inhibiting as a result of steric hindrance of the reactive grouping[271].

Schubert and Sanders drew the conclusion, based on the overall scheme of inhibitory activities of the model substances investigated, that the antibacterial action of sugar solutions given high energy radiation is due to the presence of α,β-unsaturated carbonyl derivatives, formed by radiolysis of the sugar in the aqueous medium. The spectrometric properties of the irradiated sugar solutions also point to the presence of such compounds. Thus, sucrose, glucose, and fructose, after irradiation in aqueous solution, gave absorption maxima which, at pH 3, lie in the region of 250 nm–260 nm and which, with increasing pH, undergo a bathochromic shift to the region of 280 nm–290 nm, with an accompanying sharp rise in extinction coefficient. The spectral properties were similar to those of methylglyoxal, and the presence in principle of a similar chromophoric system was considered likely (scheme 5.8).

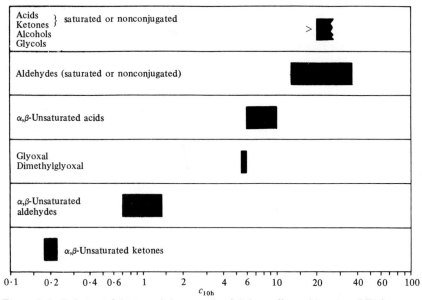

Figure 5.6. Relative inhibition of the growth of *Salmonella typhimurium* LT2 by different classes of compounds, expressed in terms of c_{10h}, that is, the concentration of inhibitor at which the initial doubling time, T_i, equals 10 h (according to Schubert and Sanders[268]).

The absorption spectra of the 2,4-dinitrophenylhydrazones in alkaline solution exhibit two absorption maxima (at 430 nm and 515 nm), markedly differing from the spectra of 2,4-dinitrophenylosazones of α-dicarbonyl compounds, which have maxima in the region 550 nm–570 nm. Although the light absorption suggests a saturated carbonyl linkage, the relative intensities of the maxima were believed to be characteristic for unsaturated carbonyl derivatives. Thus the authors conclude, from the combined microbiological and spectrometric data, that sugar solutions, on high energy irradiation, form α,β-unsaturated carbonyl compounds by radiolysis. These compounds are formed by dehydration of dicarbonyl compounds, the primary products of the irradiation. For glucose scheme 5.9 was postulated. On autoclaving, the equilibrium would be shifted in favour of the cytotoxic products, and this would explain the microbiological results.

It is apparent that this interpretation of the microbiological inhibitory effects of irradiated sugar solutions is highly speculative, and requires the isolation and unambiguous identification of the inhibitory substances. Schubert and Sanders[268] have achieved this and report on the constitution of the cytotoxic α,β-unsaturated carbonyl compounds obtained from irradiated D-glucose, D-mannose, and D-galactose. These carbonyl compounds are responsible for the microbiological inhibitory effects, and

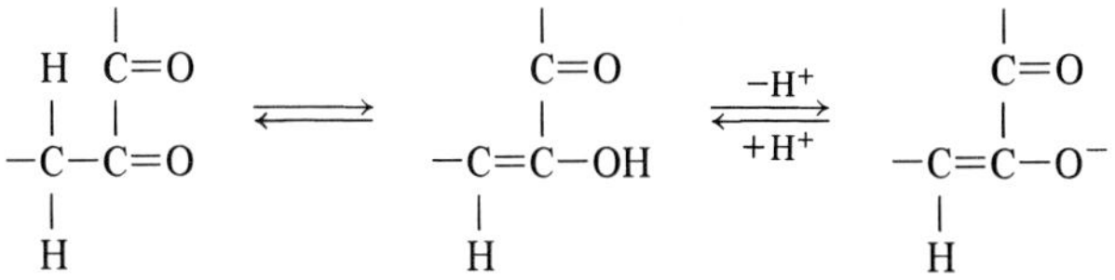

$$\begin{array}{c} | \\ H-C=O \\ | \quad | \\ -C-C=O \\ | \\ H \end{array} \rightleftharpoons \begin{array}{c} | \\ C=O \\ | \\ -C=C-OH \\ | \\ H \end{array} \underset{+H^+}{\overset{-H^+}{\rightleftharpoons}} \begin{array}{c} | \\ C=O \\ | \\ -C=C-O^- \\ | \\ H \end{array}$$

Scheme 5.8

glucose ↓ hν

$$\begin{array}{c} H-C=O \\ | \\ C-OH \\ \| \\ C-H \\ | \\ H-C-OH \\ | \\ C=O \\ | \\ CH_2OH \end{array} \underset{+H_2O}{\overset{-H_2O}{\rightleftharpoons}} \begin{array}{c} H-C=O \\ | \\ H-C-OH \\ | \\ HO-C-H \\ | \\ H-C-OH \\ | \\ C=O \\ | \\ CH_2OH \end{array} \underset{+H_2O}{\overset{-H_2O}{\rightleftharpoons}} \begin{array}{c} H-C=O \\ | \\ H-C-OH \\ | \\ C-H \\ \| \\ C-OH \\ | \\ C=O \\ | \\ CH_2OH \end{array} \rightleftharpoons \begin{array}{c} H-C=O \\ | \\ H-C-OH \\ | \\ H-C-H \\ | \\ C=O \\ | \\ C-OH \\ | \\ CHOH \end{array}$$

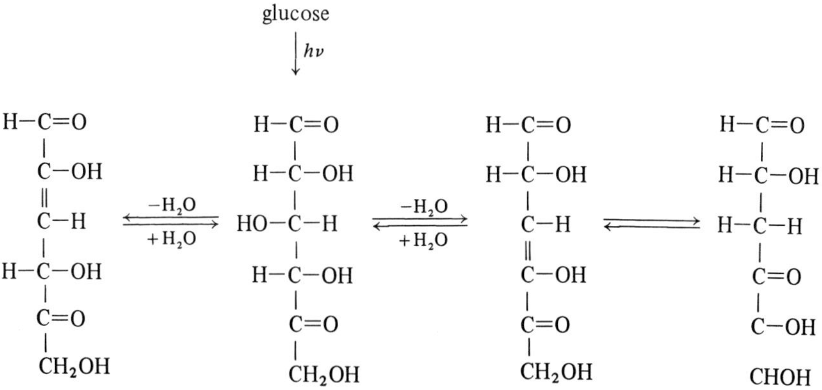

Scheme 5.9

for the chemical and spectrometric properties of the original irradiated solutions. J. Schubert (personal communication) has in the meantime been successful in determining unambiguously by mass spectroscopy and synthesis the structure of the active substance obtained from irradiated glucose solution.

High energy irradiation of 6-deoxy sugars (D-fructose and D-rhamnose) also gives carbonyl compounds, for which α-ketoaldehyde structures have been deduced from the light absorption of the 2,4-dinitrophenylosazones in alkaline solution.

Schubert et al.[272] have also examined the action of model carbonyl compounds and irradiated sugar solutions, using human lymphocytes as the test system *in vitro*. It was shown that α,β-unsaturated carbonyl compounds, namely crotonaldehyde and ethyl vinyl ketone, have greater cytotoxic activity than saturated aldehydes, such as glyceraldehyde, glycolaldehyde, and propanal.

Thus the minimal concentrations found for inhibition of mitosis were 0·8–3 mM for saturated aldehydes, but only 0·1–0·3 mM for α,β-unsaturated aldehydes. Further, the concentrations needed to achieve chromosome breakages are approximately 0·3 mM for saturated, and 0·08 mM for unsaturated, aldehydes. With increasing concentration the aldehydes effect ever greater damage to the chromosomes until a complete disintegration takes place. No rearrangements of damaged chromosomes have been observed, and it was concluded that protective mechanisms responsible for the reaggregation of damaged chromosomes are destroyed. It is interesting to note that the range of concentrations in which the morphological damage to the chromosome occurs is remarkably narrow. Half the concentration of crotonaldehyde or ethyl vinyl ketone which is effective in producing chromosome breakages or disintegration is virtually without effect. Such extraordinarily steep dose–response curves are very characteristic of α,β-unsaturated aldehydes (see the discussion of the biological effects of the 4-hydroxyenals; section 3.2). The cytotoxic effects of the irradiated aqueous solutions of hexoses and of autoclaved 2-deoxyhexoses and 2-deoxyriboses correspond to those of model α,β-unsaturated carbonyl compounds.

The authors also tackled the question of the mechanism of the biological effects of α,β-unsaturated carbonyl compounds. They suspect that the cytotoxic effects are overwhelmingly based on 1,4-addition of sulphydryl groups to the conjugated unsaturated system, thereby giving stable thioethers. Alternatively, but to a smaller extent, a direct addition may take place to the carbonyl group with formation of less stable mercaptals. Finally, one must also consider reactions with the amino groups of proteins and nucleic acids, which give rise to substituted amines (by addition to the conjugated system) or Schiff's bases (by direct addition).

The hypothesis for the mechanism of reaction of α,β-unsaturated carbonyls with sulphydryl groups is quite plausible, but has not yet been experimentally verified. The experimental proof of the reaction with biologically functional sulphydryl groups, and of the mechanism of the reaction, is discussed in detail in the section on 4-hydroxyenals (section 3.2).

The conclusions of Schubert et al. appear to be substantiated by the data obtained by Lange et al. for glyceraldehyde (see section 4.1) and by the results quoted in section 5.1.4. Thus, in irradiated glucose solution, α,β-unsaturated carbonyl compounds are formed which may then react with amino groups, and in particular with those of nucleic acids.

References
1. Whipple, E. B., 1970, *J. Am. Chem. Soc.*, **92**, 7183.
2. Hopwood, D., 1969, *Histochemie*, **20**, 127.
3. Silverstein, R. M., Bassler, G. C., 1967, *Spectrometric Identification of Organic Compounds* (John Wiley, New York).
4. Colowick, S., Lazarow, A., Racker, E., Schwarz, D. R., Stadtman, E., Waelsch, H., 1954, *Glutathione* (Academic Press, New York), p.171.
5. Kuhn, E., 1950, *J. Biol. Chem.*, **187**, 289.
6. Königstein, J., Federonko, M., 1970, *Proc. Anal. Chem. Conf. 3rd*, **2**, 113.
7. Vail, S. L., Moran, C. M., Barker, R. H., 1965, *J. Org. Chem.*, **30**, 1195.
8. Edwards, J. M., Weiss, U., Gilardi, R. D., Karle, J. L., 1968, *Chem. Commun.*, **24**, 1649.
9. Kliegman, J. M., Barnes, R. K., 1970, *J. Heterocycl. Chem.*, **7**, 1153.
10. Kliegman, J. M., Barnes, R. K., 1970, *J. Org. Chem.*, **35**, 3140.
11. Davis, P., Tabor, B. E., 1963, *J. Polymer Sci. Part A*, **1**, 799.
12. Maurer, K., Woltersdorf, E. H., 1938, *Z. Physiol. Chem.*, **254**, 18.
13. Johnson, D. P., Critchfield, F. E., Ruch, J. E., 1962, *Anal. Chem.*, **34**, 1389.
14. Underwood, G. E., Siem, R. A., Gerpheide, S. E., 1959, *Proc. Soc. Exp. Biol. Med.*, **100**, 312.
15. Bengelsdorf, I., 1953, *J. Am. Chem. Soc.*, **75**, 3138.
16. Bowes, J. H., Cater, C. W., 1968, *Biochim. Biophys. Acta*, **168**, 341.
17. Takahashi, K., 1968, *J. Biol. Chem.*, **243**, 6171.
18. Schubert, M. P., 1936, *J. Biol. Chem.*, **114**, 341.
19. Kermack, W. O., Matheson, N. A., 1957, *Biochem. J.*, **65**, 48.
20. Cliffe, E. E., Waley, S. G., 1961, *Biochem. J.*, **79**, 475.
21. Vince, R., Wadd, W. B., 1969, *Biochem. Biophys. Res. Commun.*, **35**, 593.
22. Nakaya, K., Horinishi, H., Shibata, K., 1967, *J. Biochem. (Tokyo)*, **61**, 345.
23. Nishimura, Y., Makino, H., Takenaka, O., Inada, Y., 1971, *Biochim. Biophys. Acta*, **227**, 171.
24. Suzuki, S., Hachimori, Y., 1968, *Nippon Kagaku Zasshi*, **89**, 614.
25. Takahashi, K., 1970, *J. Biochem. (Tokyo)*, **68**, 659.
26. Keil, B., 1971, *FEBS Lett.*, **14**, 181.
27. Kotaki, A., Harada, M., Nagi, K., 1968, *J. Biochem. (Tokyo)*, **64**, 537.
28. Fukui, T., Kamogaoa, A., Nikuni, Z., 1970, *J. Biochem. (Tokyo)*, **67**, 211.
29. Glazer, A. N., 1970, *Annu. Rev. Biochem.*, **39**, 101.
30. Tabor, B. E., 1968, *J. Appl. Polym. Sci.*, **12**, 1967.
31. Buttkus, H., 1968, *J. Am. Oil Chem. Soc.*, **46**, 88.
32. Tannenbaum, S. R., Barth, H., Le Roux, J. P., 1969, *J. Agr. Food Chem.*, **17**, 1353.
33. Milch, R. A., 1965, *South. Med. J.*, **58**, 153.
34. Staehelin, M., 1959, *Biochim. Biophys. Acta*, **31**, 448.

35 Shapiro, R., Hachman, J., 1966, *Biochemistry*, **5**, 2799.
36 Shapiro, R., Cohen, J. B., Shiney, S. J., 1969, *Biochemistry*, **8**, 238.
37 Nakaya, K., Takenaka, O., Horinishi, H., Shibata, K., 1968, *Biochim. Biophys. Acta*, **161**, 23.
38 Litt, M., Hancock, V., 1967, *Biochemistry*, **6**, 1848.
39 Litt, M., 1969, *Biochemistry*, **8**, 3249.
40 Broude, N. E., Budovskii, E. J., Kochetkov, N. K., 1967, *Mol. Biol. (USSR)*, **1**, 214.
41 Brooks, B. R., Klamerth, O. L., 1968, *Eur. J. Biochem.*, **5**, 178.
42 Klamerth, O. L., 1968, *Biochim. Biophys. Acta*, **155**, 271.
43 Kisch, B., 1932, *Biochem. Z.*, **253**, 373.
44 Kuhn, E., 1950, *J. Biol. Chem.*, **187**, 289.
45 Spolter, P. D., Adelman, R. C., Weinhouse, S., 1965, *J. Biol. Chem.*, **240**, 1327.
46 Kuhn, E., 1952, *J. Biol. Chem.*, **194**, 603.
47 Horton, A. A., Packer, L., 1970, *J. Gerontol.*, **25**, 199.
48 Klamerth, O. L., 1968, *Biochim. Biophys. Acta*, **155**, 271.
49 Guidotti, G. G., Loretti, L., Ciaranfi, E., 1965, *Eur. J. Cancer*, **1**, 23.
50 Ruffo, A., Adinolfi, A., Budillon, G., Capobianco, G., 1962, *Biochem. J.*, **85**, 593.
51 Szent-Györgyi, A., 1960, *An Introduction to Subcellular Biology* (Academic Press, New York), p.125.
52 Együd, L., 1965, *Proc. Nat. Acad. Sci. USA*, **54**, 200.
53 Együd, L., Szent-Györgyi, A., 1966, *Proc. Nat. Acad. Sci. USA*, **56**, 203.
54 Együd, L., 1967, *Curr. Mod. Biol.*, **1**, 14.
55 Schubert, M. P., 1935, *J. Biol. Chem.*, **111**, 671.
56 Együd, L., 1968, *Curr. Mod. Biol.*, **2**, 128.
57 Scaife, J., 1969, *Experientia*, **25**, 178.
58 Ratner, S., Clarke, H., 1937, *J. Am. Chem. Soc.*, **59**, 200.
59 Együd, L., Szent-Györgyi, A., 1968, *Science*, **160**, 1140.
60 Otsuka, H., Együd, L., 1967, *Cancer Res.*, **27**, 1498.
61 Greeg, C. T., 1968, *Exp. Cell Res.*, **50**, 65.
62 Szent-Györgyi, A., Együd, L., McLaughlin, J., 1967, *Science*, **155**, 539.
63 Szent-Györgyi, A., 1968, *Perspect. Biol. Med.*, **11**, 350.
64 Knock, F. E., 1966, *J. Am. Geriatr. Soc.*, **14**, 883.
65 Vescia, A., Romano, M., Cerra, M., 1964, *Boll. Soc. Ital. Biol. Sper.*, **40**, 2047.
66 Szent-Györgyi, A., 1967, *Proc. Nat. Acad. Sci. USA*, **57**, 1642.
67 Otsuka, H., Együd, L., 1968, *Biochim. Biophys. Acta*, **165**, 172.
68 Jellum, G., 1968, *Biochim. Biophys. Acta*, **170**, 430.
69 Apple, M. A., Greenberg, D. M., 1967, *Cancer Chemother. Rep.*, **51**, 455.
70 Kenny, C., Sparkes, B., 1968, *Science*, **161**, 1344.
71 Sparkes, B., Kenny, C., 1969, *Proc. Nat. Acad. Sci. USA*, **64**, 920.
72 Sparkes, B., Kenny, C., 1970, *Int. Z. Klin. Pharmakol. Ther. Toxicol.*, **3**, 223.
73 Doerr, W., 1948, *Verh. Dtsch. Ges. Pathol.*, **32**, 278.
74 Doerr, W., 1948, *Naturwissenschaften*, **35**, 125.
75 Helge, H., 1958, *Verh. Dtsch. Ges. Pathol.*, **42**, 158.
76 Ling, N. R., 1961, *Br. J. Haematol.*, **7**, 299.
77 Procaccini, R. L., de Fanti, D. R., de Feo, J. I., 1971, *Biochim. Biophys. Res. Commun.*, **43**, 484.
78 Pogo, B. G. T., Allfrey, V., Mirsky, A., 1966, *Proc. Nat. Acad. Sci. USA*, **55**, 805.
79 Herbage, D., Reynier, M., Palloc, D., 1969, *C. R. Soc. Biol.*, **163**, 2634.
80 Palloc, D., Reynier, M., Venot, C., Regli, P., 1969, *C. R. Soc. Biol.*, **163**, 2630.
81 Ashwood-Smith, M. J., Robinson, D. M., Barnes, J. H., Bridges, B. A., 1967, *Nature (London)*, **216**, 137.
82 Ashwood-Smith, M. J., Barnes, J. H., Huckle, J., Bridges, B. A., 1969, *Radiat. Prot. Sensitation, Proc. Int. Symp. 2nd*, p.183.

83 Barnes, J. H., Ashwood-Smith, M. J., Bridges, B. A., 1969, *Int. J. Radiat. Biol.*, **15**, 285.
84 Underwood, G. E., Weed, S. D., 1956, *Proc. Soc. Exp. Biol. Med.*, **93**, 421.
85 French, F. A., Freedlander, B. L., 1958, *Cancer Res.*, **18**, 172.
86 Együd, L., Szent-Györgyi, A., 1968, *Science*, **180**, 1940.
87 Jerzykowski, T., Matusewsky, W., Otronsek, N., Winter, R., 1970, *Neoplasma*, **17**, 25.
88 Skipper, H. E., Schabel, F. M., Wilcox, W. S., 1964, *Cancer Chemother. Rep.*, **35**, 1.
89 Knock, F. E., 1966, *Lancet*, **9**, 824.
90 Stewart, T. D., Donnally, L. H., 1932, *J. Am. Chem. Soc.*, **54**, 3559.
91 Kruger, P. G., 1955, *Radiat. Res.*, **3**, 1.
92 Vince, R., Wadd, W. B., 1969, *Biochem. Biophys. Res. Commun.*, **35**, 593.
93 Vince, R., Daluge, S., 1971, *J. Med. Chem.*, **14**, 35.
94 Jellum, E., Elgjo, K., 1970, *Biochem. Biophys. Res. Commun.*, **38**, 575.
95 Issekutz, B., 1969, *The Chemotherapy of Cancer* (Academiai Kiado, Budapest), p.17.
96 Lewis, K. F., Majane, E. H., Weinhouse, S., 1959, *Cancer Res.*, **19**, 97.
97 Hopkins, F. G., Morgan, E. J., 1952, *Biochem. J.*, **194**, 119.
98 Monder, C., 1967, *J. Biol. Chem.*, **242**, 4603.
99 Jellum, E., 1968, *Biochim. Biophys. Acta*, **165**, 375.
100 Schmidt, A., 1931, *C. R. Soc. Biol.*, **108**, 152.
101 Bloch, H., Lehr, H., Erlenmeyer, H., Vogler, K., 1945, *Helv. Chim. Acta*, **28**, 1411.
102 Jalander, W. Y., 1936, *Arch. Exp. Pathol. Pharmacol.*, **180**, 628.
103 Baumann, E., 1938, *Klin. Wochenschr.*, **17**, 382.
104 Kulshrestha, D. C., Marth, E. H., 1970, *J. Milk Food Technol.*, **33**, 305.
105 German, A., Panouse-Perrin, J., Zekam, J., Dause, M. J., 1967, *Ann. Pharm. Fr.*, **25**, 783.
106 Good, H., 1971, *Patent; Chem. Abstr.*, **75**, 121428.
107 Cooper, R. A., Anderson, A., 1970, *FEBS Lett.*, **11**, 273.
108 Freedberg, W. B., Kistler, W. S., Lin, E. C., 1971, *J. Bacteriol.*, **108**, 137.
109 Barrett, P. A., Beveridge, E., Bradley, P. L., Brown, C. G., Bushby, S. R., Clarke, M. L., Neal, R. A., Smith, R., Wilde, J. K., 1965, *Nature (London)*, **206**, 1340.
110 Tiffany, B. D., Wright, J. B., Moffet, R. B., Heinzelman, R. V., 1957, *J. Am. Chem. Soc.*, **79**, 1682.
111 Moffet, R. B., Tiffany, B. D., Asperen, B. D., Heinzelman, R. V., 1957, *J. Am. Chem. Soc.*, **79**, 1687.
112 Wright, J. B., Lincoln, E. H., Heinzelman, R. V., 1957, *J. Am. Chem. Soc.*, **79**, 1690.
113 Underwood, G. E., Siem, R. A., Gerpheide, S. A., 1959, *Proc. Soc. Exp. Biol. Med.*, **100**, 312.
114 Underwood, G. E., Weed, S. D., 1961, *Virology*, **13**, 138.
115 Hartman, F. W., Kelly, A. R., Grippo, G. A., 1955, *Gastroenterology*, **28**, 244.
116 De Bock, C. A., Brug, J., Walop, N. J., 1957, *Nature (London)*, **179**, 706.
117 Cavallini, G., 1963, *Int. Symp. Chemother. 2nd, Naples* 1961, Part 2, p.36.
118 Muecke, H., Sproessig, M., 1967, *Arch. Exp. Veterinärmed.*, **21**, 307.
119 Remis, H. E., 1969, *Progr. Antimicrobial Anticancer Chemother. Proc. Int. Congr. Chemother. 6th*, 1969, Part 2, p.844.
120 Sproessig, M., Muecke, H., 1963, *Int. Symp. Chemother. 2nd, Naples* 1961, Part 2, p.318.
121 Sproessig, M., Muecke, H., 1963, *Acta Virol.*, **7**, 472.

122 Patton, S., Keeney, M., Kurtz, G. W., 1951, *J. Am. Oil. Chem. Soc.*, **28**, 319.
123 Dahle, L. K., Hill, E. G., Hollmann, R. T., 1962, *Arch. Biochem. Biophys.*, **98**, 253.
124 Sinnhuber, R. O., Yu, T. C., 1958, *Food Res.*, **23**, 626.
125 Kwon, T. W., Olcott, H. S., 1966, *Nature (London)*, **210**, 214.
126 Saslaw, L. D., Anderson, H. J., Waravdekar, W. S., 1963, *Nature (London)*, **200**, 1098.
127 Saslaw, L. D., Waravdekar, W. S., 1965, *Radiat. Res.*, **24**, 375.
128 Sawicki, E., Stanley, T. W., Johnson, H., 1963, *Anal. Chem.*, **35**, 199.
129 Thiculin, G., Morre, J., Richou, L., 1963, *Ann. Nutr. Aliment.*, **17**, 385.
130 Kwon, T. W., Olcott, H. S., 1966, *J. Food Sci.*, **31**, 552.
131 Täufel, K., Zimmermann, R., 1962, *Z. Lebensm. Unters. u. Forsch.*, **118**, 105.
132 Scherz, H., 1968, *Experientia*, **24**, 420.
133 Ambe, K., Tappel, A., 1961, *J. Food Sci.*, **26**, 448.
134 Morre, J., Morrazani-Pelletier, S., 1966, *C. R. Acad. Sci.*, **262**, 1729.
135 Scherz, H., Stehlik, G., Bancher, E., Kaindl, K., 1967, *Mikrochim. Acta*, **5**, 915.
136 Wertheim, J. H., Proctor, B. E., 1956, *J. Dairy Sci.*, **39**, 391.
137 Morre, J., Richou, L., 1963, *Bull. Acad. Vet. Fr.*, **36**, 349.
138 Barber, A. A., Bernheim, F., 1967, *Advances in Gerontological Research*, Volume II, Ed. B. L. Strehler (Academic Press, New York).
139 Paeker, L., Deamer, D. W., Heath, R. L., 1967, *Advances in Gerontological Research*, Volume II, Ed. B. L. Strehler (Academic Press, New York).
140 Haugaard, N., 1968, *Physiol. Rev.*, **48**, 311.
141 Harman, D., 1972, *J. Am. Geriatr. Soc.*, **20**, 145.
142 Kwon, T. W., Brown, W. D., 1965, *Fed. Proc., Fed. Amer. Soc. Exp. Biol.*, **24**, 592.
143 Crawford, D. L., Yu, T. C., Sinnhuber, R. O., 1966, *J. Agric. Food Chem.*, **14**, 182.
144 Brooks, B. R., Klamerth, O. L., 1968, *Eur. J. Biochem.*, **5**, 178.
145 Placer, Z., Veselkova, A., Rath, R., 1965, *Experientia*, **21**, 19.
146 Horton, A. A., Packer, L., 1970, *J. Gerontol.*, **25**, 199.
147 Stocks, J., Dormandy, D. L., 1971, *Brit. J. Haematol.*, **20**, 95.
148 Jose, P. J., Slater, T. F., 1972, *Proc. Biochem. Soc.*, 1972, 141P.
149 Ikegami, F., 1970, *Juzen Igakkai Zasshi*, **79**, 610.
150 Takada, A., Ikegami, F., 1970, *Lab. Invest.*, **23**, 421.
151 Klages, F., 1949, *Ber. Dtsch. Chem. Ges.*, **82**, 374.
152 Chio, K. S., Tappel, A. L., 1969, *Biochemistry*, **8**, 2821.
153 Harman, D., 1957, *J. Gerontol.*, **12**, 199.
154 Strehler, B. L., Mildvan, A. S., 1962, *Biological Aspects of Aging*, Ed. N. W. Shock (Columbia University Press, New York), p.174.
155 Hendley, D. D., Mildvan, A. S., Reporter, M. C., Strehler, B. L., 1963, *J. Gerontol.*, **18**, 144.
156 Hyden, H., Lindström, B., 1950, *Discuss. Faraday Soc.*, **9**, 436.
157 Chio, K. S., Tappel, A. L., 1969, *Biochemistry*, **8**, 2827.
158 Friedman, M., Cavins, J. F., Wall, J. S., 1965, *J. Am. Chem. Soc.*, **87**, 3672.
159 Lawley, P. D., 1962, *The Molecular Basis of Neoplasia* (University of Texas Press, Texas), p.123.
160 Mashio, F., Kimura, Y., 1960, *Nippon Kagaku Zasshi*, **81**, 434.
161 Saunders, J., May, J. R. K., 1963, *Chem. Ind. (London)*, 1355.
162 Kwon, T. W., Watts, B. M., 1964, *J. Food Sci.*, **29**, 294.
163 Bacon, N., George, W. O., Stringer, B. H., 1965, *Chem. Ind. (London)*, 1377.
164 Kwon, T. W., Menzel, D. B., Olcott, H. S., 1965, *J. Food Sci.*, **30**, 808.

165 Hüttel, R., 1941, *Ber. Dtsch. Chem. Ges.*, **74**, 1825.
166 Kwon, T. W., 1966, *J. Chromatogr.*, **24**, 193.
167 Protopova, T. V., Skoldinov, A. P., 1958, *J. Gen. Chem. USSR*, **28**, 241.
168 Klamerth, O. L., Levinsky, H., 1969, *FEBS Lett.*, **3**, 205.
169 Horton, A. A., 1970, *Biochem. J.*, **116**, 19P.
170 Crawford, D. L., Yu, T. C., Sinnhuber, R. O., 1967, *J. Food Sci.*, **32**, 332.
171 Landsberg, J. D., Sinnhuber, R. O., 1965, *J. Am. Oil Chem. Soc.*, **42**, 821.
172 Taggart, J., 1971, *Biochem. J.*, **123**, 4P.
173 Ewart, J. A. D., 1968, *J. Sci. Food Agric.*, **19**, 370.
174 Buttkus, H., 1968, *J. Am. Oil Chem. Soc.*, **46**, 88.
175 Tannenbaum, S. R., Barth, H., Le Roux, J. P., 1969, *J. Agr. Food Chem.*, **17**, 1353.
176 Buttkus, H., 1967, *J. Food Sci.*, **32**, 432.
177 Stork, G., Brizzolara, A., Landesman, H., Szmuszkovicz, J., Terrel, R., 1963, *J. Am. Chem. Soc.*, **85**, 207.
178 Hirs, C. H. W., Moore, S., Stein, W. H., 1956, *J. Biol. Chem.*, **219**, 623.
179 Hirs, C. H. W., 1962, *Brookhaven Symp. Biol.*, **15**, 154.
180 Menzel, D. B., 1967, *Lipids*, **2**, 83.
181 Holtkamp, D. E., Hill, R. M., 1951, *Arch. Biochem. Biophys.*, **34**, 216.
182 Recknagel, R. O., Ghoshal, A. K., 1965, *Fed. Proc., Fed. Amer. Soc. Exp. Biol.*, **24**, 299.
183 Deitrich, R. A., 1966, *Biochem. Pharmacol.*, **15**, 1911.
184 Crawford, D. L., Sinnhuber, R. O., Stout, F. M., Oldfield, J. E., Kaufmes, J., 1965, *Toxicol. Appl. Pharmacol.*, **7**, 826.
185 Quash, G., Taylor, D. R., 1970, *Clin. Chim. Acta*, **30**, 17.
186 Abraham, K., 1968, *Eur. J. Biochem.*, **5**, 143.
187 Moruzzi, G., Barbiroli, B., Calderera, C., 1968, *Biochem. J.*, **107**, 609.
188 Torrielli, M. V., Ugazio, G., 1970, *Life Sci.*, **9**, 1.
189 Aso, C., Aito, Y., 1962, *Makromol. Chem.*, **58**, 195.
190 Yokota, K., Suzuki, Y., Ishii, Y., 1965, *Kogyo Kagaku Zasshi*, **68**, 2459; *Chem. Abstr.*, **65**, 13835.
191 Richards, F. M., Knowles, J. R., 1968, *J. Mol. Biol.*, **37**, 231.
192 Hardy, P. M., Nichols, A. C., Rydon, H. N., 1969, *Chem. Commun.*, 565.
193 Milch, R. A., 1964, *Biochim. Biophys. Acta*, **93**, 45.
194 Anderson, P. J., 1967, *J. Histochem. Cytochem.*, **15**, 652.
195 Hopwood, D., 1967, *Histochemie*, **11**, 289.
196 Fregerio, N. A., Shaw, M. J., 1969, *J. Histochem. Cytochem.*, **17**, 176.
197 Hopwood, D., 1969, *Histochemie*, **20**, 127.
198 Trelstad, L. R., 1969, *J. Histochem. Cytochem.*, **17**, 756.
199 Habeeb, A. F. S. A., 1968, *Arch. Biochem. Biophys.*, **126**, 16.
200 Hopwood, D., Allen, C. R., McCabe, M., 1970, *Histochem. J.*, **2**, 137.
201 Munton, T. J., Russel, A. D., 1970, *J. Appl. Bacteriol.*, **33**, 410.
202 Bowes, J. H., Cater, C. W., 1968, *Biochim. Biophys. Acta*, **168**, 341.
203 Robinson, I. D., 1964, *J. Appl. Polym. Sci.*, **8**, 1903.
204 Grant, R. A., 1965, *Biochem. J.*, **97**, 5C.
205 Milch, R. A., 1965, *Nature (London)*, **205**, 1108.
206 Cater, C. W., 1965, *J. Soc. Leather Trades Chem.*, **49**, 455.
207 Tabor, B. E., 1968, *J. Appl. Polym. Sci.*, **12**, 1967.
208 Milch, R. A., Clifford, R. E., Murray, R. A., 1966, *Nature (London)*, **210**, 1042.
209 Milch, R. A., Frisco, L. J., Szymkoviak, E. A., 1965, *Biorheology*, **3**, 9.
210 Moczar, M., Moczar, E., Payran, P., 1969, *C. R. Acad. Sci.*, **268**, 2734.
211 Hopwood, D., 1969, *Histochemie*, **17**, 151.

212 Bowes, J. H., Cater, C. W., 1966, *J. Roy. Micros. Soc.*, **85**, 193.
213 Ewart, J. A. D., 1968, *J. Sci. Food Agr.*, **14**, 370.
214 Bowes, J. H., Cater, C. W., 1965, *J. Appl. Chem.*, **15**, 296.
215 Quiochio, F. A., Richards, F. M., 1966, *Biochemistry*, **5**, 4062.
216 Filachione, E. M., Leorn, A. H., Ard, J. S., 1967, *J. Amer. Leather Chem. Assoc.*, **62**, 450.
217 Wold, F., 1967, *Methods Enzymol.*, **88**, 617.
218 Quiochio, F. A., Richards, F. M., 1964, *Proc. Nat. Acad. Sci. USA*, **52**, 833.
219 Ching-Wang, J. H., Tu, J. J., 1969, *Biochemistry*, **8**, 4403.
220 Sabatini, D. S., Bensch, K., Barrnett, R. J., 1963, *J. Cell. Biol.*, **17**, 19.
221 Habeeb, A. F. S. A., 1954, *J. Immunol.*, **102**, 457.
222 Reichlin, M., Schmurre, J. J., 1968, *Proc. Soc. Exp. Biol. Med.*, **128**, 347.
223 Avrameas, S., 1969, *Immunochemistry*, **6**, 43.
224 Sachs, D. H., Winn, H. J., 1970, *Immunochemistry*, **7**, 581.
225 Avrameas, S., Ternynck, T. H., 1969, *Immunochemistry*, **6**, 53.
226 Ling, N. R., 1961, *Br. J. Haematol.*, **7**, 299.
227 Boyer, J. T., 1967, *Nature (London)*, **214**, 291.
228 Bring, D. H., Weyand, J. G. M., Stavitsky, A. B., 1967, *Proc. Soc. Exp. Biol. Med.*, **124**, 1166.
229 Onkelinx, E., Meuldermans, W., Jonian, M., Lontie, R., 1969, *Immunology*, **16**, 35.
230 Reale, E., Luciano, L., 1970, *Histochemie*, **23**, 144.
231 Fahimi, H. D., Drochmans, P., 1965, *J. Micros. (Oxford)*, **4**, 737.
232 Deamer, D. W., Utsumin, K., Packer, L., 1967, *Arch. Biochem. Biophys.*, **121**, 641.
233 Utsumi, K., Packer, L., 1967, *Arch. Biochem. Biophys.*, **121**, 633.
234 Packer, L., Wrigglesworth, J. M., Fortes, P. A., Pressman, B. C., 1968, *J. Cell Biol.*, **39**, 382.
235 Manti, K. E., 1970, *Plant Physiol.*, **45**, 563.
236 Packer, L., Allen, J. M., Starks, M., 1968, *Arch. Biochem. Biophys.*, **128**, 142.
237 Yoshida Yoshio, 1969, *Plant Cell Physiol.*, **10**, 555.
238 Packer, L., Greville, G. D., 1969, *FEBS Lett.*, **3**, 112.
239 Horton, A. A., Packer, L., 1970, *J. Gerontol.*, **25**, 199.
240 Horton, A. A., Packer, L., 1970, *Biochem. J.*, **116**, 19P.
241 Behrman, J., 1966, *Clin. Chem.*, **12**, 211.
242 Morel, F., Baker, R. F., Weyland, H., Knust-Graichen, P. W., 1971, *J. Cell Biol.*, **48**, 91.
243 Hasselbach, W., Fiehn, W., Makinose, M., Migala, A. J., 1968, *Molecular Basis of Membrane Function, Symp.*, *1968*, 299.
244 Pohl, S. L., Birnbaumer, L., Rodbell, M., 1971, *J. Biol. Chem.*, **246**, 1849.
245 Koshiba Kimikazu, 1968, *Acta Med. Okayama*, **22**, 11.
246 Hughes, R. C., Thurman, P. F., 1970, *Biochem. J.*, **119**, 925.
247 Ross, P. W., 1960, *J. Clin. Pathol.*, **19**, 318.
248 Krzywicka, H., 1970, *Rocz. Panstw. Zakl. Hig.*, **21**, 87.
249 Pepper, R. E., Chandler, V. L., 1963, *Appl. Microbiol.*, **11**, 384.
250 Stonehill, A. A., Krop, S., Borick, P. M., 1963, *Am. J. Hosp. Pharm.*, **20**, 458.
251 Snyder, R. W., Cheatle, E. L., 1965, *Am. J. Hosp. Pharm.*, **22**, 321.
252 Rubbo, S. D., Gardner, J. F., Webb, R. L., 1967, *J. Appl. Bacteriol.*, **30**, 78.
253 Sidwell, R. W., Westbrook, L., Dixon, G. J., Happich, W. F., 1970, *Appl. Microbiol.*, **19**, 53.
254 Bachrach, U., Rosenkovitch, E., 1972, *Appl. Microbiol.*, **23**, 232.
255 Blough, H. A., 1966, *J. Bacteriol.*, **92**, 266.
256 Graham, J. L., Jaeger, R. F., 1968, *Appl. Microbiol.*, **16**, 177.
257 Bachrach, U., Don, S., Wiener, H., 1971, *J. Gen. Virol.*, **13**, 415.

258 Wojciak Zofia, 1969, *Rocz. Panstw. Zakl. Hig.*, **20**, 119.
259 Sabel, F. L., Hellman, A., McDade, J. J., 1969, *Appl. Microbiol.*, **17**, 645.
260 Kremzner, T. L., Harter, D. H., 1970, *Biochem. Pharmacol.*, **19**, 2541.
261 German, A., Panouse-Perrin, J., Zekam, J., Dause, M. J., 1967, *Ann. Pharm. Fr.*, **25**, 783.
262 Sabel, F. L., Hellman, A., McDade, J. J., 1969, *Appl. Microbiol.*, **17**, 645.
263 Borik, P. M., 1965, *Biotechnol. Bioeng.*, **7**, 435.
264 Rubbo, S. D., Gardner, J. F., Webb, R. C., 1967, *J. Appl. Bacteriol.*, **30**, 78.
265 Munton, T. J., Russel, A. D., 1970, *J. Appl. Bacteriol.*, **33**, 410.
266 Buttkus, H., 1968, *J. Am. Oil Chem. Soc.*, **46**, 88.
267 Buttkus, H., 1972, *J. Am. Oil Chem. Soc.*, **49**, 613.
268 Schubert, J., Sanders, E., 1971, *Nature (London), New Biol.*, **233**, 199.
269 Holsten, R., Sugii, M., Stewart, F., 1965, *Nature (London)*, **208**, 850
  Philips, G., 1968, *Energetics and Mechanisms in Radiation Biol.*, Acad. Prep., N.4, 131.
270 Schubert, J., 1970, *J. Gen. Microbiol.*, **64**, 37.
271 Gutsche, C., 1967, *Chemistry of Carbonyl Compounds* (Prentice Hall, Englewood Cliffs, NJ).
  Royals, E., 1954, *Advanced Organic Chem.*, (Prentice Hall, Englewood Cliffs, NJ).
272 Schubert, J., Pan, Y., Wald, N., 1971, *Environmental Mutagens. Soc.*, second annual meeting, abstr.

Stable derivatives of aldehydes

In this chapter, only a brief survey is attempted of the action of stable derivatives of aldehydes on bacteria, fungi, viruses, and tumour cells. In this context the most frequently examined derivatives of aldehydes have been the thiosemicarbazones, guanylhydrazones, and various Schiff's bases. The carbon–nitrogen double bond in these derivatives is so stable that as a rule no dissociation to the free aldehyde takes place under physiological conditions. The biochemical effects of these compounds are therefore conditioned by their special chemical constitution and not by the aldehyde possibly set free as an intermediate. Only in the case of a few Schiff's bases has it been shown that they could dissociate into free aldehyde and amine at pH 7·0 in an aqueous medium, and the biological activity (for example, antitumour action) is then decreased proportionately to the increase in the velocity of hydrolysis of the Schiff's base[1].

6.1 Antibacterial and fungistatic effects

Domagk et al.[2] were the first to report, in 1946, on the activity of p-aminobenzaldehyde thiosemicarbazone against tuberculosis bacilli. Subsequent investigations showed that many thiosemicarbazones of aliphatic, aromatic, and heterocyclic aldehydes have tuberculostatic[3-7] and general bacteriostatic[8-10] properties. Barrett et al.[11] in 1965 described the antibacterial activity of a large number of bisthiosemicarbazones of α-ketoaldehydes, among which the glyoxal derivatives proved to be the most active. Significant bacteriostatic activity was also shown by the Schiff's bases of aniline and substituted anilines (4-chloroaniline, 4-thiocyanato-aniline) with aromatic aldehydes (hydroxybenzaldehyde, 3,5-dibromosalicyl-aldehyde, nicotinic acid esters of hydroxybenzaldehyde)[12,13].

Many authors have reported on the fungistatic activity of the thiosemi-carbazones of aliphatic and aromatic aldehydes[14-17] and of five-membered heterocyclic aldehydes[18]. Schiff's bases of 4-thiocyanatoaniline with aromatic aldehydes also have fungicidal properties[19].

6.2 Antiviral effects

Thiosemicarbazones of isatin, namely isatin-3-thiosemicarbazone and 1-methylisatin-3-thiosemicarbazone (the antiviral drug Marboran ® or Methisazone ®), have been used with good effect against viral infections in animals and man: pox virus, adenovirus, and herpes virus groups[20-24].
The concentrations necessary for an inhibition of virus multiplication are not toxic for the host cells[25]. The mechanism of action was also examined; in the case of pox virus it was supposed that the thiosemicarbazones act only in the late phase of the replication cycle and principally prevent the maturation of the viruses[25-27].

6.3 Carcinostatic effects

Brockman et al.[28], in 1956, reported on the antileukaemia effect of the thiosemicarbazone of pyridine-2-carboxaldehyde. These findings were taken up in 1963 by French et al., who examined in considerable detail numerous thiosemicarbazones of pyridine- and isoquinoline-aldehydes[29-33]. The most reactive substances were found to be 5-hydroxypyridine-2-carboxaldehyde thiosemicarbazone, 3-hydroxypyridine-2-carboxaldehyde thiosemicarbazone, and isoquinoline-1-carboxaldehyde thiosemicarbazone. These compounds inhibit *in vitro* a number of tumours in mice, such as L-1210 leukaemia, sarcoma-180 (ascites), L-5178 Y lymphoma, Levis lung carcinoma, and Ehrlich ascites carcinoma. The carcinostatic actions of various isoquinoline-1-carboxaldehyde thiosemicarbazones were also intensively examined by Agrawal and Sartorelli[34-38,65]. These authors made the important discovery that isoquinoline-1-carboxaldehyde thiosemicarbazone prevented the conversion in tumour cells of ribonucleotides into deoxyribonucleotides, and thereby effected a strong inhibition of DNA synthesis. On the other hand, RNA and protein syntheses were influenced only slightly. The ribonucleotide diphosphate reductase from mammalian tumours contains in the active centre an iron ion (nonhaemin Fe^{2+}), which is essential for the activity. French et al. first postulated[39], and later proved[40], that thiosemicarbazones, especially those of pyridine-2-carboxaldehyde and isoquinoline-1-carboxaldehyde, form stable complexes with ferrous ions (Fe^{2+}). They have also discussed the mechanism of the antitumour action[32] and believe that the thiosemicarbazones inhibit the ribonucleotide reductase by formation of a complex with ferrous ion, and that this is the actual cause of the antitumour effect.

pyridine-2-carboxaldehyde thiosemicarbazone

isoquinoline-1-carboxaldehyde thiosemicarbazone

A different group of aldehyde derivatives with antitumour activity is derived from α-ketoaldehydes. The first results were obtained in 1958 by Freedlander and French[41], who found that the bisguanylhydrazones of glyoxal and methylglyoxal increased the survival time of mice with leukaemia L 1210. Since then, many research workers have examined this group of compounds, the most active being bisthiosemicarbazones, bis-4-alkylthiosemicarbazones, and bisguanylhydrazones of glyoxal, methylglyoxal, and kethoxal (table 6.1). These compounds were administered either intraperitoneally, or orally with the food. The inhibition of the tumour growth can be intensified in animal experiments by feeding with a diet deficient in vitamin B_6[42-44].

Table 6.1. Stable aldehyde and ketone derivatives with antitumour activity. TSC = thiosemicarbazone; GH = guanylhydrazone.

Aldehyde	Derivative	Active against	References
Glyoxal	bis-GH	Leukaemia L 1210	41
Glyoxal	bis-TSC	Sarcoma-180	39
Glyoxal	bis-alkyl-GH	Sarcoma-180	39
Methylglyoxal	bis-GH	Leukaemia L 1210	41
Methylglyoxal	bis-methyl-TSC	Sarcoma-180, Walker 256	39, 41, 51
Kethoxal	bis-TSC	Sarcoma-180, Walker 256, leukaemia L 1210, Jensen sarcoma, lymphosarcoma	47, 51-54
Kethoxal	bis-methyl-TSC	Sarcoma-180	52
4-Hydroxy-2-butenone	TSC	Levis-King carcinoma	55
1,4- and 1,5-dicarbonyl compounds	bis-alkyl-TSC	Sarcoma-180	56
Polyaldehydes (oxidized starch)	Poly-TSC	Sarcoma-180	57
Pyridin-2-carboxaldehyde	TSC	Leukaemia L 1210, sarcoma-180, Levis-King carcinoma, etc.	28-31, 33
Isoquinolin-1-carboxaldehyde	TSC	Leukaemia L 1210, sarcoma-180, Levis-King carcinoma, etc.	34-38, 58, 59
Benzaldehyde N-mustards	TSC and hydrazones	Walker 256, leukaemia L 1210	60-63
Benzaldehyde N-mustards	Schiff's-bases with arylamines	Walker 256, leukaemia L 1210, Dunmung leukaemia	50, 64
Salicylaldehyde and other aromatic aldehydes	Schiff's-bases with arylamines	Leukaemia L 1210	1

The cytotoxic action of α-ketoaldehyde bisthiosemicarbazones and bisguanylhydrazones is probably connected with their ability to form complexes with metal ions, especially cupric (Cu^{2+}) and zinc (Zn^{2+}) ions[39,45,46]. It was assumed that in the tumour the metal ion balance is disturbed, but the precise mechanism is not known.

The most interesting derivative appears to be kethoxal bisthiosemicarbazone, which in animal experiments is very active against many tumours (Walker carcinosarcoma-256, sarcoma-180, Murphy–Sturm lymphosarcoma, leukaemia L 1210, spontaneous adenocarcinoma, and many others). Pharmacological studies[47] with mice, rats, dogs, and monkeys showed, however, that this compound produces severe, acute, and chronic side effects in the mammalian organism, so that clinical application is very limited.

A French research group[48] reported in 1970 on the successful clinical use of methylglyoxal bisguanylhydrazone in leukaemia. The adducts of methylglyoxal with simple amines, which exhibit antitumour activity, have been described by Szent-Györgyi and were referred to in section 5.1.5.2.

In recent years a multitude of thiosemicarbazones of aliphatic, aromatic, and heterocyclic aldehydes, and also Schiff's bases, have been prepared and shown to be active against animal tumours. A selection of the most important of these derivatives is given in table 6.1. A few of these derivatives have also been tested clinically[49,50].

References
1 Hodnett, E. M., Tai, J., 1971, *J. Med. Chem.*, **14**, 1115.
2 Domagk, G., Benisch, R., Mietzsch, F., Schmidt, H., 1946, *Naturwissenschaften*, **33**, 315.
3 Anderson, F. E., Duca, C. J., Scudi, J. V., 1951, *J. Am. Chem. Soc.*, **73**, 4961.
4 Bernstein, J., Yale, H. L., Losee, K., Holsing, M., Martins, J., Lott, W. A., 1951, *J. Am. Chem. Soc.*, **73**, 906.
5 Hagenbach, R. E., Gusin, H., 1952, *Experientia*, **8**, 184.
6 Caldwell, H. C., Nobles, W. L., 1956, *J. Am. Pharm. Assoc. Sci. Ed.*, **45**, 279.
7 Dodgen, D. F., Nobles, W. L., 1957, *J. Am. Pharm. Assoc. Sci. Ed.*, **46**, 437.
8 Bernstein, J., Yale, H. L., Losee, K., Holsing, H., Martins, J., Lott, W. A., 1951, *J. Am. Chem. Soc.*, **73**, 906.
9 Weller, L. E., Sell, H. M., Gottshall, R. Y., 1954, *J. Am. Chem. Soc.*, **76**, 1959.
10 Nardi, D., Massarani, E., Tajana, A., Degen, L., Magistretti, M. J., 1967, *J. Med. Chem.*, **10**, 530.
11 Barret, P. A., Beveridge, E., Bradley, P. L., Brown, C. G., Bushby, S. R., Clarke, M. L., Neal, R. A., Smith, R., Wilde, J. K., 1965, *Nature (London)*, **206**, 1340.
12 Weuffen, W., Pohlouden-Fabini, R., 1966, *Arch. Pharm.*, **299**, 777.
13 Weuffen, W., Theus, P. M., 1967, *Pharmazie*, **22**, 428.
14 Benns, R. G., Gingras, B. A., Bayley, C. H., 1960, *Appl. Microbiol.*, **8**, 353.
15 Gingras, B. A., Colin, G., Bayley, C. H., 1965, *J. Pharm. Sci.*, **54**, 1674.
16 Wiles, D. M., Gingras, B. A., Suprunchuk, T., 1967, *Can. J. Chem.*, **45**, 1375.
17 Wiles, D. M., Suprunchuk, T., 1969, *J. Med. Chem.*, **12**, 526.
18 Wiles, D. M., Suprunchuk, T., 1971, *J. Med. Chem.*, **14**, 252.
19 Pohloudek-Fabini, R., Weuffen, W., 1964, *Arch. Pharm.*, **297**, 554.
20 Hamre, D., Bernstein, J., Donovick, R., 1950, *Proc. Soc. Exp. Biol. Med.*, **73**, 275.
21 Thompson, R. L., Minton, S. A., Officer, J. E., Hitchings, G. H., 1953, *J. Immunol.*, **70**, 229.

22 Turner, W., Bauer, D. J., Nimmo-Smith, R. H., 1962, *Br. Med. J.*, **i**, 1317.
23 Bauer, D. J., 1965, *Ann. N. Y. Acad. Sci.*, **130**, 110.
24 Munro, T. W., Sabina, C. R., 1970, *J. Gen. Virol.*, **7**, 55.
25 Bauer, D. J., Apostolon, K., 1966, *Science*, **154**, 796.
26 Appleyard, G., Hume, V. B., Westwood, J. C., 1965, *Ann. N. Y. Acad. Sci.*, **130**, 92.
27 Magee, W. E., Bach, M. V., 1965, *Ann. N. Y. Acad. Sci.*, **130**, 80.
28 Brockman, R. W., Thompson, J. R., Bell, M. J., Skipper, H. E., 1956, *Cancer Res.*, **16**, 167.
29 French, F. A., Blanz, E. J., 1966, *Cancer Res.*, **26**, 1638.
30 French, F. A., Blanz, E. J., 1966, *J. Med. Chem.*, **9**, 585.
31 Blanz, E. J., French, F. A., 1968, *Cancer Res.*, **28**, 2419.
32 French, F. A., Blanz, E. J., Do Amaral, J. R., French, D. A., 1970, *J. Med. Chem.*, **13**, 1117.
33 Blanz, E. J., French, F. A., Do Amaral, J. R., French, D. A., 1970, *J. Med. Chem.*, **13**, 1124.
34 Sartorelli, A. C., 1967, *Biochem. Biophys. Res. Commun.*, **27**, 26.
35 Sartorelli, A. C., 1967, *Pharmacologist*, **9**, 192.
36 Agrawal, K. C., Booth, B. A., Sartorelli, A. C., 1968, *J. Med. Chem.*, **11**, 700.
37 Agrawal, K. C., Sartorelli, A. C., 1968, *J. Pharm. Sci.*, **57**, 1948.
38 Agrawal, K. C., Sartorelli, A. C., 1969, *J. Med. Chem.*, **12**, 171.
39 French, F. A., Freedlander, B. L., 1958, *Cancer Res.*, **18**, 1290.
40 French, F. A., Lewis, A. E., Blanz, E. J., Sheena, A. H., 1965, *Fed. Proc. Fed. Am. Soc. Exp. Biol.*, **24**, 402.
41 Freedlander, B. L., French, F. A., 1958, *Cancer Res.*, **18**, 360.
42 Mihich, E., Nichol, C. A., 1963, *Proc. Am. Assoc. Cancer Res.*, **4**, 44.
43 Petering, H. G., Buskirk, H. H., Crim, A. J., 1963, *Proc. Am. Assoc. Cancer Res.*, **4**, 52.
44 Mihich, E., Nichol, C. A., 1965, *Cancer Res.*, **25**, 794.
45 Petering, H. G., Buskirk, H. H., Kupiecki, F. P., 1965, *Fed. Proc. Fed. Am. Soc. Exp. Biol.*, **24**, 454.
46 Mihich, E., Mulhern, A. I., 1965, *Fed. Proc. Fed. Am. Soc. Exp. Biol.*, **24**, 454.
47 Mihich, E., Simpson, C. L., Mulhern, A. I., 1965, *Cancer Res.*, **25**, 1417.
48 Groupe Coopérateur des Leucémies et Hématosarcomes, 1970, *Eur. J. Cancer*, **6**, 57.
49 Goldin, A., Serpick, A. A., Mantel, N., 1966, *Cancer Chemother. Rep.*, **50**, 190.
50 Popp, F. D., 1964, *J. Med. Chem.*, **7**, 210.
51 Mihich, E., Nichol, C. A., 1965, *Cancer Res.*, **25**, 1410.
52 French, F. A., Freedlander, B. L., 1960, *Cancer Res.*, **20**, 505, Suppl. VII.
53 Petering, H. G., Buskirk, H. H., Underwood, G. E., 1964, *Cancer Res.*, **24**, 367.
54 Sartorelli, A. C., Booth, B. A., 1967, *Cancer Res.*, **27**, 1614.
55 Prescott, B., 1967, *J. Med. Chem.*, **10**, 484.
56 Barry, V. C., Conalty, M. L., McCormick, J. E., McElhinney, R. S., McInerney, M. R., O'Sullivan, J. F., 1970, *J. Med. Chem.*, **13**, 421.
57 Barry, V. C., Conalty, M. L., McCormick, J. E., McElhinney, R. S., O'Sullivan, J. F., 1966, *Proc. R. Ir. Acad. Sect. B*, **64**, 335.
58 French, F. A., Blanz, E. J., 1965, *Cancer Res.*, **25**, 1454.
59 Agrawal, K. C., Cushley, R. J., McMurray, W. J., Sartorelli, A. C., 1970, *J. Med. Chem.*, **13**, 431.
60 Florvall, L., 1970, *Acta Pharm. Suec.*, **7**, 87.
61 Popp, F. D., 1962, *J. Med. Pharm. Chem.*, **5**, 627.
62 Elderfield, R. C., Roy, J., 1967, *J. Med. Chem.*, **10**, 918.
63 Martinez, A. P., Lee, W. W., 1967, *J. Med. Chem.*, **10**, 1192.
64 Modi, J. D., Sabnis, S. S., Deliwala, C. V., 1970, *J. Med. Chem.*, **13**, 935.
65 Sartorelli, A. C., Agrawal, K. C., Moore, E. C., 1971, *Biochem. Pharmacol.*, **20**, 3119.

Part 2

Naturally occurring aldehydes and their biological functions

Schematic survey of the most important aldehydes involved in intermediary metabolism

7.1 Aliphatic long chain aldehydes from animal tissues and from serum
Saturated and unsaturated lipid aldehydes have been found in rat, ox, and dog heart[1,2], in serum[3], and in liver, kidney, skeletal muscle, and brain[4]. Under the analytical conditions selected only the free aldehydes were determined, and not those which could be set free through acidic hydrolysis of plasmologens. The lipid aldehyde fraction of heart consists of approximately 90% saturated aldehydes, that is predominantly hexa- and octadecanal with smaller amounts of tetra-, penta-, and heptadecanals. Also present are small quantities of unsaturated aldehydes with chain lengths of fifteen to eighteen carbon atoms. It is interesting that no polyunsaturated aldehydes were found, whereas polyunsaturated fatty acids were always present in the lipid fraction. These latter are of considerable biological significance. Hardly anything is known about the origin and significance of these fatty aldehydes. The label in [^{14}C]acetate that had been applied intraperitoneally to rats, and in [^{14}C]palmitate which, bound to albumin, had been infused into dogs, could be recovered in the fatty aldehydes. It can therefore be supposed that these are synthesized *in vivo*, possibly with the involvement of acyl-CoA reductases, such as have already been isolated from *Brassica oleracaea*[5], *Euglena gracilis*[6], and *Clostridium butyricum*[7]. In any case palmityl-CoA is reduced to the corresponding aldehyde, with involvement of NAD, by a rat brain homogenate[8]. Tietz *et al.*[9] found that rat liver metabolized batyl alcohol to glycerol and a fatty acid, and that an aldehyde appeared as an intermediate in this reaction. This may be an alternative for the origin of fatty aldehydes.

In mammalian organisms enzymes which split the 1-O-alk-1-enylglycerol ether linkage of the plasmologens[10,11] were found. In view of the wide distribution of plasmologens, Wood and Healy[12] have suggested that this enzymatic activity could be the source of the fatty aldehydes. It is also argued by Spener and Mangold[13] that these free aldehydes are artefacts caused by autohydrolysis or chemical hydrolysis of the plasmologens, especially because of the structural analogy between the free aldehydes and those linked to glycerol as 1-enol ethers. Other authors, on the other hand, find no particular correspondence between the distribution patterns of free, and of bound, aldehydes[14,2].

Gilbertson *et al.*[1,2] have also demonstrated that the activity of [^{14}C] acetate administered to rats appeared in free fatty aldehydes as well as in 1-O-alk-1-enylglycerol. In addition experimental findings exist which suggest that 1-*O*-alk-1-enyl ethers are formed in a reaction of the corresponding aldehyde with the hydroxyl group of the glycerol precursor, followed by a dehydration[15]. We are far from being able to draw precise conclusions about the physiological significance of free fatty aldehydes.

7.2 α-Ketoaldehydes

7.2.1 Glyoxal

Glyoxal has been found to be present in various plant products such as tobacco leaves[17] and tomatoes[18], and to occur in many alcoholic beverages[19,20] and in milk[21], but glyoxal does not appear to occur physiologically in mammalian organisms.

As shown by many authors[23-25], glyoxal is formed on γ-irradiation of aqueous solutions of glucose, fructose, or sucrose, and is therefore always present in foodstuffs containing carbohydrate which have been irradiated. The presence of glyoxal, like that of malonaldehyde, has thus often been used to indicate that an irradiation has taken place. As shown by Cobb and Day[25], glyoxal may be produced together with other aldehydes and dicarbonyl compounds through autoxidation of unsaturated lipids.

7.2.2 Methylglyoxal

Methylglyoxal occurs in the flavouring matter of tomatoes[18], in wine[19], in beer[40,43], and in oxidized fats[25]. It can also be found in mammalian urine[44], blood[45], or milk[46] under special circumstances, such as in vitamin B_1 deficiency.

It has long been known that small quantities of methylglyoxal may be formed on incubation of tissue homogenates with glucose or other substrates[26]. Since the discovery of glyoxalase[27,28], which converts methylglyoxal into lactate, it had been assumed that methylglyoxal-lactate forms a pathway secondary to the classical Embden–Meyerhof degradation. However, this hypothesis has been discarded for a variety of reasons; a detailed discussion of this is given by Lewis *et al.*[29] and by Riddle and Lorenz[30]. When it became known that methylglyoxal is easily and spontaneously formed from trioses, the question arose whether in general methylglyoxal, which is always present in animal tissue, has an enzymatic or nonenzymatic origin. Walton and McLeon[31] are even of the opinion that the earlier findings on the occurrence of methylglyoxal in animal tissues are artefacts conditioned by the preparative procedures. Using a double-isotope technique with [1,3-^{14}C]methylglyoxal and [3,5,6-^{3}H]2,4-dinitrophenylhydrazine, these authors were unable to find any endogenous methylglyoxal in rat liver. In the opinion of Walton and McLeon[31] the different methods of proof, which are based on the glyoxalase-1 reaction or on oxidation with alkaline hydrogen peroxide, do not distinguish between different 1,2-dicarbonyl compounds. A positive reaction with arsenophosphotungstate is also given by other aldehydes and ketones. Finally, the conversion of methylglyoxal into the bis-2,4-dinitrophenylhydrazone is possible only under such drastic reaction conditions that methylglyoxal may indeed be formed from glyceraldehyde or from dihydroxyacetone.

A series of enzymes which convert various substrates into methylglyoxal has been detected. It is not known, however, whether these reactions are in fact of significance *in vivo*. Various authors have shown that methylglyoxal is formed in mammalian tissues, via the aminoacetone cycle, from glycine and threonine with the participation of acetyl-CoA[32-34].

In some tissues methylglyoxal may also be formed enzymatically from acetoacetate[35], lactaldehyde[36], and aminopropanol[37]. An enzyme which converts dihydroxyacetone phosphate, derived from the degradation of glucose or from glycerol, into methylglyoxal appears to be present in *Escherichia coli*. The methylglyoxal is then further metabolized to lactate and pyruvate. A pathway secondary to the sequence occurring in normal glycolysis may be present in this instance. Nothing is known, however, about the physiological significance for bacteria of this possible secondary pathway[38,39].

As early as 1937 Needham and Lehman[40] had pointed out that nonenzymatic conversion of glyceraldehyde into methylglyoxal was a possibility. In 1968 Riddle and Lorenz[30] showed that the formation of methylglyoxal from glyceraldehyde and dihydroxyacetone takes place at physiological pH and temperature, and that the reaction is catalysed by inorganic phosphate, tris buffer, arsenite, and other polyvalent anions.

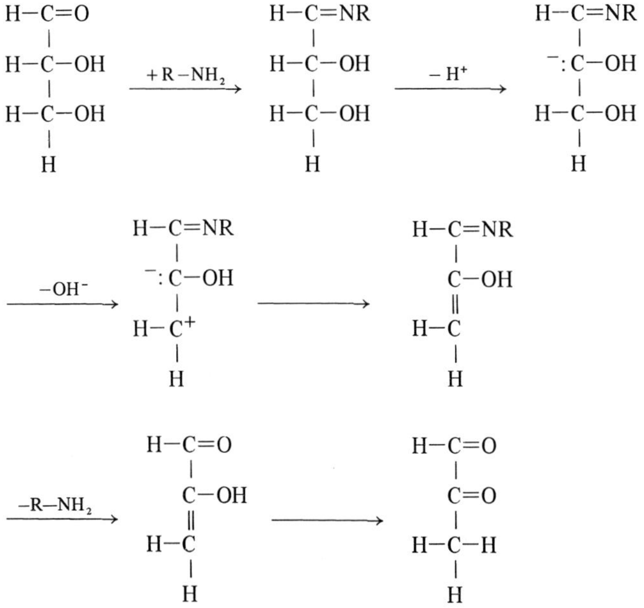

$$\begin{array}{c} H-C=O \\ | \\ H-C-OH \\ | \\ H-C-OH \\ | \\ H \end{array} \xrightarrow{+R-NH_2} \begin{array}{c} H-C=NR \\ | \\ H-C-OH \\ | \\ H-C-OH \\ | \\ H \end{array} \xrightarrow{-H^+} \begin{array}{c} H-C=NR \\ | \\ ^-:C-OH \\ | \\ H-C-OH \\ | \\ H \end{array}$$

$$\xrightarrow{-OH^-} \begin{array}{c} H-C=NR \\ | \\ ^-:C-OH \\ | \\ H-C^+ \\ | \\ H \end{array} \longrightarrow \begin{array}{c} H-C=NR \\ | \\ C-OH \\ || \\ H-C \\ | \\ H \end{array}$$

$$\xrightarrow{-R-NH_2} \begin{array}{c} H-C=O \\ | \\ C-OH \\ || \\ H-C \\ | \\ H \end{array} \longrightarrow \begin{array}{c} H-C=O \\ | \\ C=O \\ | \\ H-C-H \\ | \\ H \end{array}$$

Scheme 7.2. Amine-catalysed formation of methylglyoxal from glyceraldehyde[41,42].

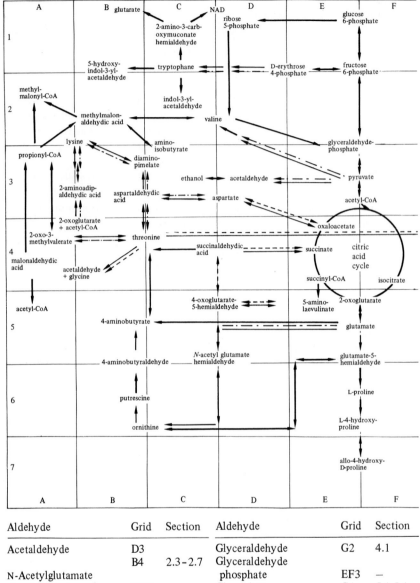

Aldehyde	Grid	Section	Aldehyde	Grid	Section
Acetaldehyde	D3		Glyceraldehyde	G2	4.1
	B4	2.3-2.7	Glyceraldehyde		
N-Acetylglutamate			phosphate	EF3	–
hemialdehyde	CD5	–	Glyoxalate	G4	5.1.5.1
2-Aminoadipate			L-4-Hydroxyglutamate		
hemialdehyde	B3	–	hemialdehyde	H6	–
4-Aminobutyraldehyde	BC5	–	N-Acetylglutamate		
2-Amino-3-carboxy-			hemialdehyde	H6	–
muconate hemialdehyde	C1	–	5-Hydroxyindol-3-yl-		
Aspartate hemialdehyde	BC3	–	acetaldehyde	B1	8.2
Betainaldehyde	H4	–	Indol-3-ylacetaldehyde	C2	8.2
Glutamate hemialdehyde	EF5	–	Isocaproicaldehyde	I5	–

Scheme 7.1. Alphabetical index of aldehydes with grid square reference and appropriate section headings.

Principal aldehydes involved in intermediary metabolism

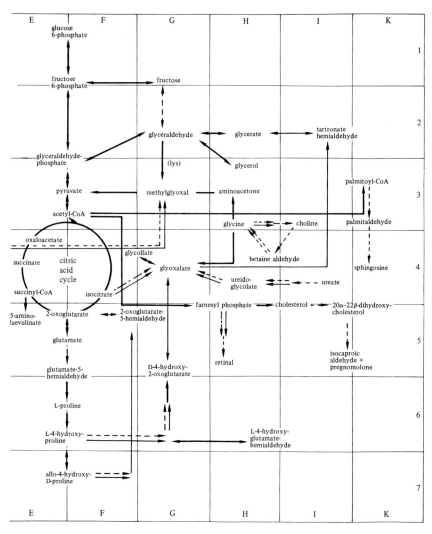

Aldehyde	Grid	Section
Malonaldehydic acid	A4	—
Methylglyoxal	G3	4.1, 5, 7.1.2
Methylmalonaldehydic acid	B2	—
2-Oxoglutarate-5-hemialdehyde	F5	—
4-Oxoglutarate-5-hemialdehyde	CD5	—
Palmitaldehyde	K3	7.1.1
Retinal	H5	8.4
Succinaldehydic acid	CD4	8.1
Tartronate hemialdehyde	J2	—

——— general pathways
——— bacteria
– – – – animals
– · – · – plants

Scheme 7.1 (continued)

These results have more recently been confirmed[41,42], and it has been further shown that lysine catalyses more strongly than phosphate and that lysine catalysis can be enhanced by phosphate. Besides lysine, amines such as ornithine, spermine, spermidine, and others also catalyse the reaction (scheme 7.2). It is assumed that a Schiff's base is formed as an intermediate which then catalysed by phosphate, splits off a molecule of water from C-2 and C-3 (cf sections 4.1 and 5).

7.2.3 Other α-ketoaldehydes

Szent-Györgyi *et al.* have attempted to show the presence of free and bound α-dicarbonyl compounds in liver. When rat liver or calf liver was treated with arsenite, 3-deoxyglucosulose (3-deoxyglucosan) was obtained (identified as the 2,4-dinitrophenylosazone, cf details in section 5.1.5).

However, it is questionable whether 3-deoxyglucosulose is actually present in the liver, since it was shown that this ketoaldehyde may be formed from glucose under the conditions used for the preparation from the liver, that is arsenite treatment[47,48]. Thus it was shown that arsenite (and also phosphate) catalyses dehydration of both glyceraldehyde and glucose, with the formation of 3-deoxyglucosulose. 3-Deoxyglucosulose may also be formed from hexoses by the Maillard reaction (scheme 7.3).

By the Maillard reaction 3-deoxyglucosulose and glucosulose are formed from glucose. Xylosulose is an intermediate in the decomposition of dehydroascorbic acid, and 3-deoxypentosulose is formed from ascorbic acid by the catalytic action of acids. Ososulose is formed from fructose–amino acids, which again are formed by an Amadori rearrangement from glucose plus amino acids.

Since 3-deoxyglucosulose inhibits cell growth even in the lowest concentrations, as shown by Szent-Györgyi *et al.* (compare section 5.4.1.5), it was supposed that it played a vital role in the regulation of cell division. However, as long as it remains uncertain whether this α-ketoaldehyde is in fact formed by the cells, it remains questionable whether it has a significance *in vivo*.

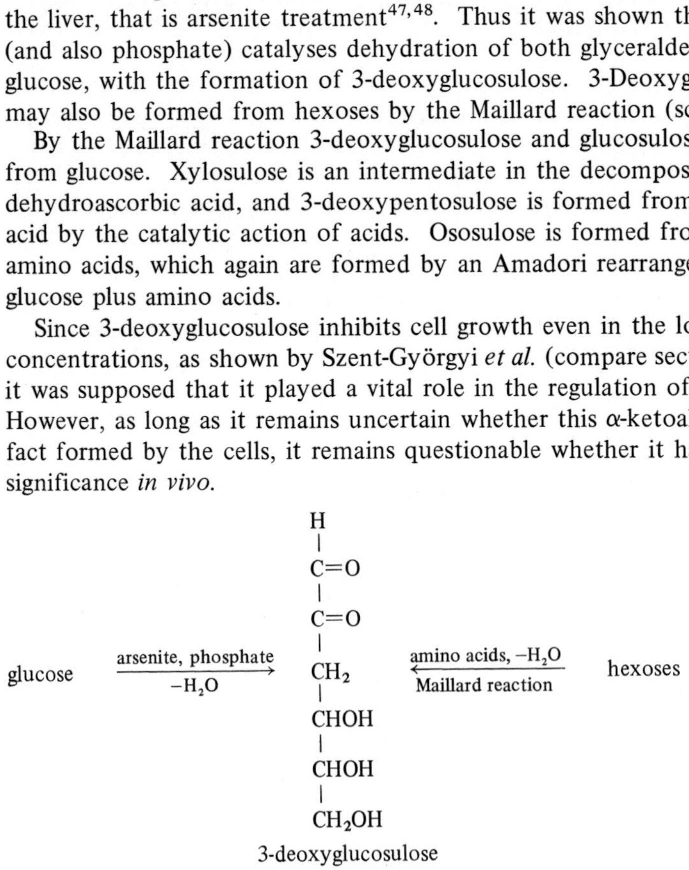

$$\text{glucose} \xrightarrow[-H_2O]{\text{arsenite, phosphate}} \begin{array}{c} H \\ | \\ C=O \\ | \\ C=O \\ | \\ CH_2 \\ | \\ CHOH \\ | \\ CHOH \\ | \\ CH_2OH \end{array} \xleftarrow[\text{Maillard reaction}]{\text{amino acids, } -H_2O} \text{hexoses}$$

3-deoxyglucosulose

Scheme 7.3

Sparkes and Kenny[49] observed that many bacteria (for example the 'less-virulent' strain of *Staphylococcus aureus*) do not grow in a nutrient medium in which human cells had been cultivated for a lengthy period. It was supposed that the cells released a substance into the medium which inhibits bacterial growth, and that this substance possibly has general significance as a growth inhibitor. Eventually the substance was identified as 4-hydroxy-2-oxobutyraldehyde[19]. It is possible that, with this aldehyde, one is also dealing with an artefact arising only during the process of isolation. Thus, Sparkes *et al.* initially isolated by means of Sephadex a lower molecular-weight, methanol-insoluble fraction, and this fraction released the inhibitor on stirring with arsenic trioxide at 37°C for twenty-four hours. If we recall the known catalytic activity of arsenite, it is conceivable that 4-hydroxy-2-oxobutyraldehyde was formed by elimination of water from a tetrose present in the nutrient medium. This, however, would leave unanswered the question of the nature of the inhibitor originally present in the nutrient medium.

The work of Kato[50] on the isolation and identification of α-ketoaldehydes from rat and calf liver is of particular interest, as these authors used conditions under which the possibility of the formation of artefacts was minimized. They deproteinized homogenates with methanol and trichloroacetic acid, and then treated directly with dinitrophenylhydrazine reagent at room temperature for thirty minutes.

By using preparative thin-layer chromatography, eighteen different α-dicarbonyl fractions were isolated. From several fractions crystalline dinitrophenylhydrazone derivatives could be obtained and identified by their infrared spectra and by comparison with synthetic materials. The following α-ketoaldehydes were identified in this manner: 3-deoxypentosulose, 3-deoxyglucosulose, xylosulose, glycosulose, and methylglyoxal. These α-ketoaldehydes are not formed in model systems consisting of ascorbic acid, glucose, valine, and phosphate under the reaction conditions employed.

According to the view of Kato *et al.* the α-ketoaldehydes found by them are metabolic intermediates of the liver and occur either free or as labile derivatives in the form of Schiff's bases or *N*-glucosides. It is known that all the α-ketoaldehydes referred to above are formed relatively easily under the catalytic action of polyvalent anions and amino acids in a nonenzymatic reaction (scheme 7.3). It is quite conceivable that such ketoaldehydes are formed also in living tissues by the Maillard reaction if reducing sugars, different amino acids, and phosphate are present together in appropriate concentrations. It is also conceivable that the Maillard reaction is partially controlled in living cells by inhibitors such as thiols.

References
1. Gilbertson, J. R., Ferrell, W. J., Gelman, R. A., 1967, *J. Lipid Res.*, **8**, 38.
2. Gilbertson, J. R., Johnson, R. C., Gelman, R. A., Buffenmyer, C., 1972, *J. Lipid Res.*, **13**, 491.
3. Gelman, R. A., Gilbertson, J. R., 1965, *Biochem. Biophys. Res. Commun.*, **20**, 427.
4. Ferrell, W. J., Radloff, J. F., Jackiw, A. B., 1969, *Lipids*, **4**, 278.
5. Kolattukudy, B. E., 1970, *Lipids*, **5**, 259.
6. Kolattukudy, B. E., 1970, *Biochemistry*, **9**, 1095.
7. Day, J. I. E., Goldfine, H., Hagen, P. O., 1970, *Biochim. Biophys. Acta*, **218**, 179.
8. Vignais, P. V., Zabin, I., 1958, *Biochem. Lipids Proc. Fifth Int. Conf. Vienna*, p.78.
9. Tietz, A., Lindberg, M., Kennedy, E. P., 1964, *J. Biol. Chem.*, **239**, 4081.
10. Barner, H. R., Lands, W. E. M., 1961, *J. Biol. Chem.*, **236**, 2404.
11. Ansell, G. B., Spanner, S., 1965, *Biochem. J.*, **94**, 252.
12. Wood, R., Healy, K., 1970, *J. Biol. Chem.*, **245**, 2640.
13. Spener, F., Mangold, H. K., 1969, *J. Lipid Res.*, **10**, 609.
14. Radloff, J. F., Ferrell, W. J., 1970, *Physiol. Chem. Phys.*, **2**, 105.
15. Piantadosi, L., Snyder, F., 1970, *Pharm. Sci.*, **59**, 283.
16. Ferrell, W. J., Radloff, J. F., 1972, *Int. J. Biochem.*, **3**, 498.
17. Rouen, H. M., 1964, *Biochemisches Taschenbuch* (Springer-Verlag, Berlin).
18. Schormüller, I., Grosch, W., 1964, *Z. Lebensm. Unters. Forsch.*, **126**, 38.
19. Suomalainen, H., Ruokainen, P., 1963, *Tek. Kem. Aikak.*, **20**, 413.
20. Palamand, S. R., Nelson, G. D., Hardwick, W. A., 1970, *Am. Soc. Brew. Chem. Proc.*, **1970**, 186.
21. Kulshrestha, D. C., Marth, E. H., 1970, *J. Milk Food Technol.*, **33**, 305.
22. Phillips, G. O., Moody, G. I., Mallock, G. L., 1958, *J. Chem. Soc.*, 3522.
23. Phillips, G. O., Moody, G. I., 1960, *J. Chem. Soc.*, 754.
24. Itzhak, G., Perikles, M., 1968, *Radiat. Res.*, **36**, 55.
25. Cobb, W. Y., Day, E. A., 1965, *J. Am. Oil Chem. Soc.*, **42**, 420, 1110.
26. Harden, A., 1923, *Alcoholic Fermentation*, 3rd edition, chapter 7 (Longmans, Green, London).
27. Neiberg, C., 1913, *Biochem. Z.*, **49**, 502.
28. Dakin, H. D., Dudley, H. W., 1913, *J. Biol. Chem.*, **14**, 155.
29. Lewis, K. F., Majane, E. H., Weinhouse, S., 1959, *Cancer Res.*, **19**, 97.
30. Riddle, V., Lorenz, F. W., 1968, *J. Biol. Chem.*, **243**, 2718.
31. Walton, D. J., McLeon, S. A., 1971, *Anal. Biochem.*, **43**, 472.
32. Elliot, H. W., 1960, *Nature (London)*, **185**, 1940.
33. Neuberger, A., Tait, G. H., 1960, *Biochim. Biophys. Acta*, **41**, 164.
34. Green, M. L., Elliott, H. W., 1964, *Biochem. J.*, **92**, 537.
35. Mulligan, L. P., Baldwin, R. L., 1967, *J. Biol. Chem.*, **242**, 1095.
36. Ting, S. M., Miller, O. N., Sellinger, D. Z., 1965, *Biochim. Biophys. Acta*, **97**, 407.
37. Higgins, I. J., Pickard, M. A., Turner, J. M., Willets, A. J., 1966, *Biochem. J.*, **99**, 278.
38. Cooper, R. A., Anderson, A., 1970, *FEBS Lett.*, **11**, 273.
39. Freedberg, W. B., Kistler, W. S., Lin, E. C., 1971, *J. Bacteriol.*, **108**, 137.
40. Needham, Y., Lehman, H., 1937, *Biochem. J.*, **31**, 1913.
41. Bonsignore, A., Leoncini, G., Ricci, D., Siri, A., 1970, *Ital. J. Biochem.*, **19**, 284.
42. Bonsignore, A., Leoncini, G., Ricci, D., Siri, A., 1972, *Ital. J. Biochem.*, **21**, 169.
43. Saha, R. B., Middlekauff, J. E., 1970, *Am. Soc. Brew. Chem. Proc.*, **1970**, 176.
44. Salem, H. M., 1955, *Arch. Biochem. Biophys.*, **57**, 20.
45. Sato, A., 1964, *Tohoku J. Exp. Med.*, **83**, 103.
46. Wako, H., 1951, *Tohoku J. Exp. Med.*, **55**, 34.

47 Otsuka, H., Egyud, L. G., 1968, *Biochim. Biophys. Acta*, **165**, 172.
48 Jellum, E., 1968, *Biochim. Biophys. Acta*, **170**, 430.
49 Sparkes, B. G., Kenny, C. P., 1970, *Int. J. Chim. Pharmacol. Ther. Toxicol.*, **3**, 233.
50 Kato, H., Tsusaka, N., Fujimaki, M., 1970, *Agric. Biol. Chem.*, **34**, 1541.
51 Kato, H., 1960, *Bull. Agric. Chem. Soc. Jn*, **24**, 1; 1962, **26**, 187.
52 Anet, E. F. L. J., 1960, *Aust. J. Chem.*, **13**, 396.
53 Reynolds, T. M., 1963, *Advances in Food Research,* Volume 12, Eds O. Chichester, E. M. Mrak, G. F. Stewart (Academic Press, New York), p.1.

8

Aldehydes with specific biological functions

8.1 Succinaldehydic acid

Succinaldehydic acid (SAA) may be formed via two metabolic pathways. The first starts from glutamic acid; this is decarboxylated to 4-aminobutyric acid which yields SAA after transamination. The other pathway is that postulated by Shemin[202]: succinyl-CoA and glycine condense to give δ-aminolaevulinic acid, which is converted into SAA via γ-oxoglutaraldehyde. An SAA dehydrogenase, dependent on NAD (NADP), reestablishes the connection of both paths with the citric acid cycle. The first pathway is the more important one with regard to the presence of SAA in brain tissue, since 4-aminobutyric acid is present there in high concentrations.

One can distinguish between effects caused by low concentrations of SAA and those caused by very high concentrations. Very small doses (1–10 mg/kg body weight), possibly corresponding to physiological concentrations, exert a dramatic slowing action on the contractility of the heart, relax smooth muscle (ileum), and lower blood pressure. Part of the hypotensive effect may be due to the bradycardiac action of the SAA dose and to the relaxation of the vascular smooth muscle. In addition, SAA shows a negative inotropic and chromotropic effect with a perfused isolated rat heart[1].

In contrast, a dose of 1 g/kg body weight raises cortical potentials[2], and a dose of 2–3 g/kg produces a calming effect on rats, resulting in drowsiness and slight hypotonia[3]. Although SAA has a calming effect, the oxygen consumption of the animals increases[2,4]. SAA at a concentration of 600–1800 mg/litre of perfusion solution causes a diminution of the amplitude of contractions and slight bradycardia in an isolated rabbit heart[5].

SAA raises motor activity and mental activity and, in certain types of schizophrenia, induces a crisis of psychomotor excitability[6]. In patients with a type of existential apraxia SAA (4–6 g, intravenously) acts on the speech centre[7].

8.2 Indol-3-ylacetaldehyde and 5-hydroxyindol-3-ylacetaldehyde

Both these aldehydes are metabolic products of tryptophan. They are produced from 5-hydroxytryptamine (serotonin) or tryptamine, respectively, by the action of a monoamine oxidase. In the brain, monoamine oxidases were found almost exclusively in the nucleolus of the locus coeruleus by means of histochemical colour techniques[8,9]. The location for the formation of 5-hydroxyindol-3-ylacetaldehyde in the brain was thus established.

Serotonin exhibits many physiological effects. One of these is the inducement of the first phase of sleep, the so-called slow wave sleep[(3)].

[(3)] Slow wave sleep is also known as light sleep, synchronized sleep, or nonrapid-eye-movement sleep.

This is followed by the next phase, paradoxical sleep [4], which is suppressed by monoamine oxidase inhibitors [10]. Jouvet[8] concluded from these findings that a deaminated serotonin metabolite must be responsible for the transition from slow wave sleep to paradoxical sleep. This hypothesis is supported by Sabelli et al.[11]. 5-Hydroxyindol-3-ylacetaldehyde and indol-3-ylacetaldehyde, like serotonin or tryptamine, produce electrophysiological effects and also effects on the behaviour of experimental animals[12].

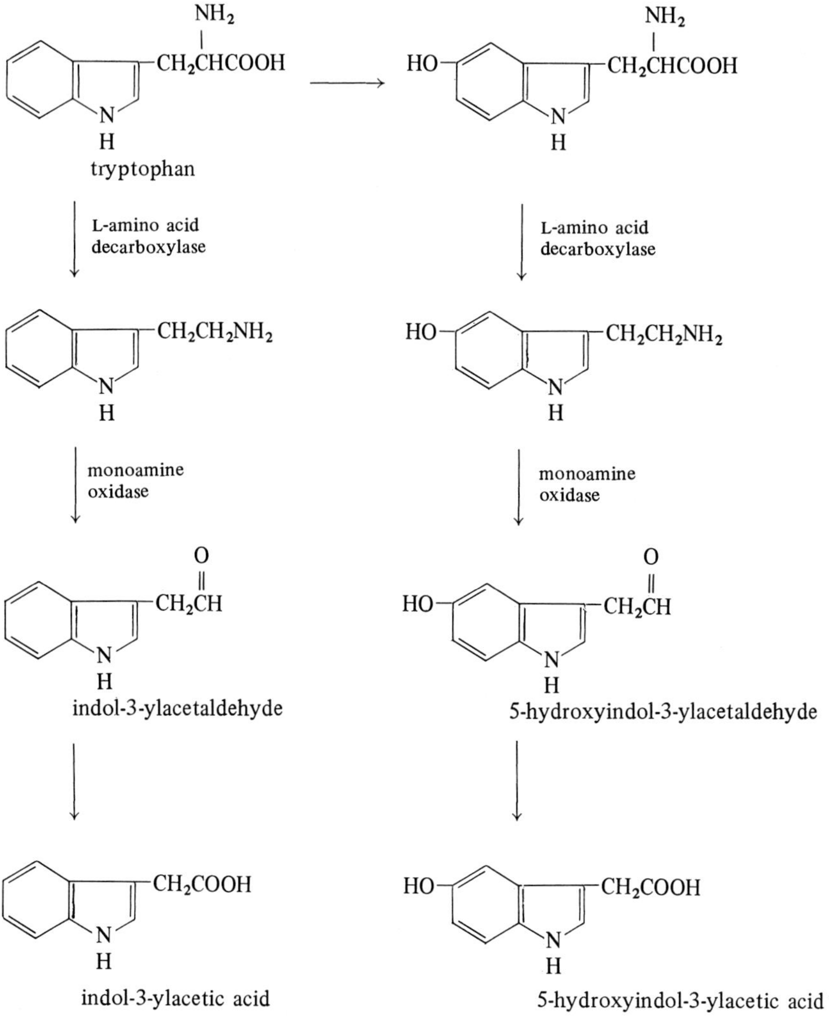

Scheme 8.1. Biosynthesis of indol-3-ylacetaldehyde and 5-hydroxyindol-3-ylacetaldehyde.

[4] Paradoxical sleep is also known as activated sleep, deep sleep, desynchronized sleep, rapid-eye-movement (REM) sleep, para sleep, or rhombencephalic sleep.

Pfeiffer *et al.*[13] obtained indirect evidence that aldehydes play a role in schizophrenia. Indol-3-ylacetaldehyde produces, in rabbits and in one- to two-day old chicks, effects on the sensory function of central neurons and on the mechanism of sleep[11].

Barbiturates, chlorpromazine, and other phenothiazine derivatives inhibit aldehyde oxidases[14] and aldehyde reductases[15]. A connection between the sedative action of the drugs mentioned and inhibition of the degradation of indol-3-ylacetaldehyde appears probable, but has not yet been proven[15].

Feldstein *et al.*[16] found that ethanol blocks the degradation of serotonin to 5-hydroxyindol-3-ylacetic acid, probably by an inhibition of the monoamine oxidase or the aldehyde dehydrogenase. The authors draw the conclusion that a possible connection exists between alcohol intoxication and altered concentrations of naturally occurring aldehydes, amines, and alcohols in the brain.

8.3 Pyridoxal–pyridoxal phosphate

If it is possible to ascribe a greater importance to one metabolite than to another, then pyridoxal must be counted among the most important aldehydes that occur in the living organism. Since this substance cannot be synthesized by the animal organism, its absence from the diet produces various deficiency symptoms, such as dermatitis, epileptic fits, hypochromic anaemia, leukopenia, and cessation of growth.

Pyridoxal derives from pyridoxol (vitamin B_6) via an enzymatically catalysed dehydrogenation. It is converted into pyridoxal phosphate with the participation of a pyridoxal kinase and ATP; dehydrogenation of pyridoxol phosphate is also possible enzymatically. The kinase reaction is reversible, but the dehydrogenase catalysed reaction is irreversible.

Pyridoxal phosphate participates as a coenzyme to animal enzymes in more than forty enzymatic reactions. In particular, many important reactions of amino acid metabolism involve pyridoxal phosphate as a coenzyme. Typical enzyme catalysed reactions which are dependent on pyridoxal phosphate are: transamination, racemization, decarboxylation, α,β- and β,γ-eliminations, β-condensation (tryptophan synthesis), and glycine (serine) condensation.

Many of these reactions are also catalysed by pyridoxal, or an analogue, in the presence of suitable polyvalent metal ions and in the absence of the apoenzyme. These nonenzymic reactions are all very unspecific and take place concurrently. The choice of suitable metal ions and a specific pH value can lead to a slight favouring of a specific reaction.

The participation of specific amino acid residues secures the binding of pyridoxal phosphate to the apoenzyme. Blocking of sulphydryl groups can completely prevent the binding of pyridoxal phosphate to the protein[17,18]. But it is most probable that these sulphydryl groups are necessary only for the maintenance of a specific structure and do not themselves participate in the binding of the coenzyme.

The protein amino groups are of much greater significance for the enzyme–coenzyme binding. Pyridoxyl-lysine could be found after sodium borohydride treatment of various enzyme–pyridoxal phosphate complexes and their enzymatic breakdown[19-28]. In this way the assumption of Jenkins and Sizer[29] that pyridoxal phosphate was bound to the apoenzyme through a Schiff's base was confirmed. Little is known, however, about the linkage between other coenzyme side groups and the apoenzyme.

The formation of a Schiff's base between substrate and coenzyme was assumed to be the first step of the reaction, on the basis of a comparison with model systems and considerations of the reaction kinetics. After reduction of the coenzyme–enzyme–substrate complex with sodium borohydride, the compound corresponding to the coenzyme–substrate–Schiff's base could also be isolated[30-33]. In this connection it is necessary to mention that the imino derivative of pyridoxal phosphate reacts very much faster with the substrate to give a Schiff's base than does the free pyridoxal phosphate. The Schiff's base binding between apoenzyme and coenzyme thus causes an activation of the coenzyme. The formation of the binding between coenzyme and substrate is catalysed via this mechanism.

Scheme 8.2 (see over) provides a summary of the mechanisms of the various pyridoxal phosphate catalysed reactions. The symptoms associated with a deficiency of pyridoxal can be ascribed largely to a disturbance of amino acid metabolism. It is suspected that the cause of epileptic seizures is a reduced γ-glutamate decarboxylase activity and a lowered γ-aminobutyric acid concentration in the brain. The first step in the biosynthesis of haemin is the pyridoxal phosphate dependent formation of δ-amino-laevulinic acid from glycine and succinyl-CoA. Since this step is hindered when pyridoxal phosphate is missing, the subsequent appearance of anaemia is readily explained.

8.3.1 Biosynthesis of pyridoxal

Only in recent years have tracer studies carried out on a *Flavobacterium* and an *Escherichia coli* mutant produced significant information. Both microorganisms incorporate ^{14}C-labelled glycerol into pyridoxol. The incorporation of the [^{14}C]glycerol by *Flavobacterium* is increased about threefold by L-leucine, L-asparagine, L-glutamine, and L-valine[34]. In the case of *E. coli* it was found that as well as glycerol, [1-^{14}C]glucose, [2,3-^{14}C]pyruvate, [3-^{14}C]aspartate, [3-^{14}C]serine, [2-^{14}C]acetate, and [2-^{14}C]leucine could also be incorporated into pyridoxol. After degradation of the vitamin B_6 molecule by the Kuhn–Roth oxidation, the labelled carbon atoms of [1,3-^{14}C]glycerol and [1-^{14}C]glucose were found in positions 2', 4', and 5'; that of [2-^{14}C]glycerol in positions 2 and 4; those of [1,3-^{14}C]pyruvate in position 2; and those of [2-^{14}C]acetate and [3-^{14}C]aspartate in position 2'. The degradation reactions carried out by Hill et al.[35] could not provide any information on positions 3, 5, or 6, and the conclusions they drew from these results were as follows: pyridoxol is

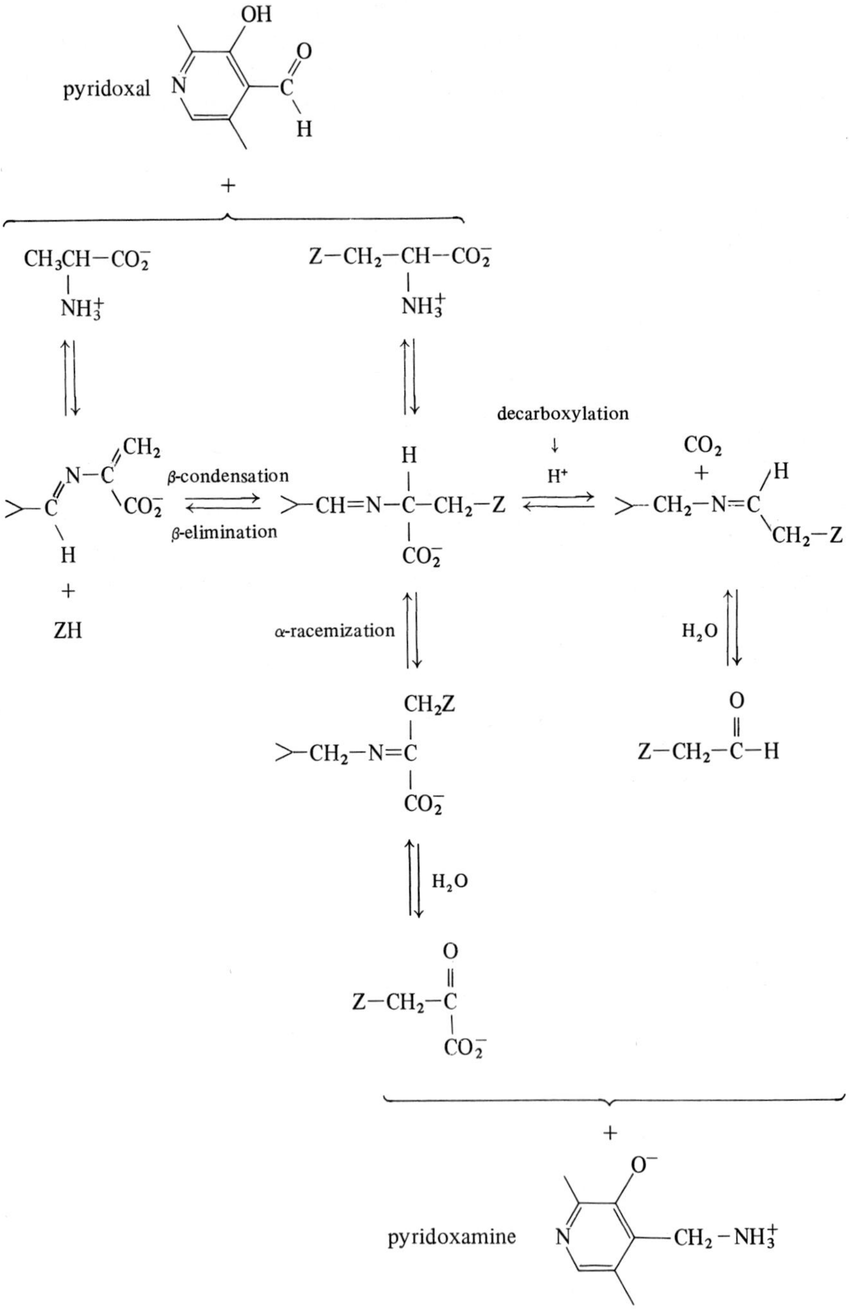

Scheme 8.2. General mechanism of pyridoxal phosphate catalysed reactions (according to Bernhard[37]).

built up by *E. coli* from three glycerol units; a three carbon atom unit is first metabolized to pyruvate, which is then incorporated as a two carbon atom unit. For mechanistic reasons the active two carbon atom unit was assumed to be acetaldehyde. The following scheme shows how the several units are distributed in the pyridoxol molecule.

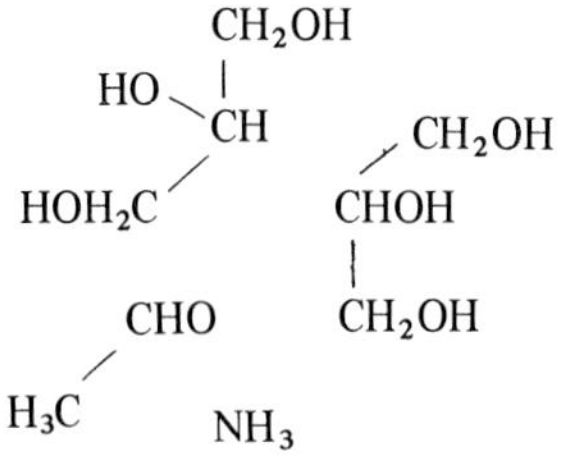

```
           CH₂OH
    HO  |
      \ CH       CH₂OH
       /       /
  HOH₂C      CHOH
              |
       CHO   CH₂OH
      /
   H₃C      NH₃
```

The first step in the biosynthesis is probably the formation of 5-deoxy-D-xylose 1-phosphate from dihydroxyacetone phosphate and acetaldehyde. This deoxy sugar then reacts with glyceraldehyde phosphate to give a branched eight carbon atom sugar. Elimination of water and phosphate, introduction of the nitrogen, and ring closure would then lead to the required end product. Complete uncertainty reigns, however, over the origin of the pyridine nitrogen atom.

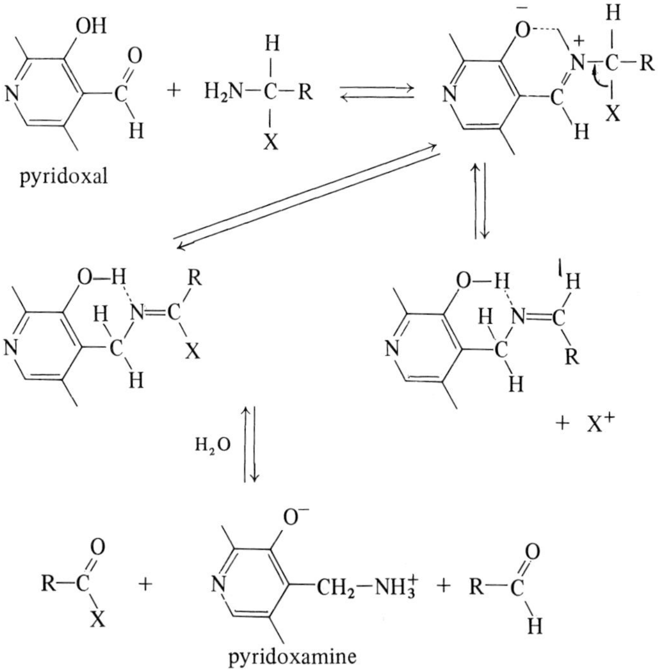

Scheme 8.2 (continued)

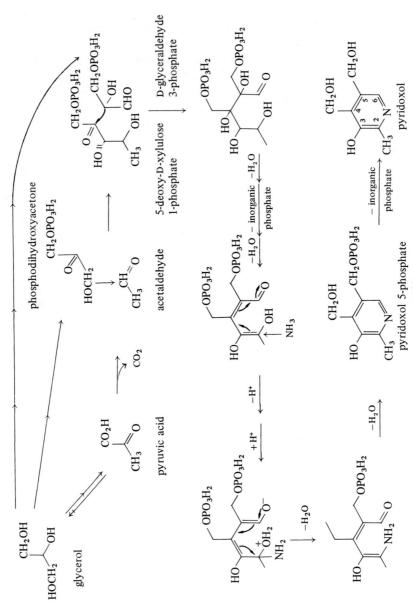

Scheme 8.3. Biosynthesis of pyridoxol (according to Hill et al.[35]).

8.4 Retinal and dehydroretinal

Retinal and 3-dehydroretinal are related to vitamin A, and therefore they have been called vitamins A_1 and A_2, respectively. Their structures are shown in scheme 8.4. They form, together with the protein component opsin, the light-sensitive pigments (rhodopsin) of the rods and cones of the retina. Since each of the two aldehydes may be the prosthetic group of the opsin from the rods or cones, one can differentiate between four main groups of visual pigment:

rod opsin + retinal = rhodopsin
cone opsin + retinal = iodopsin
rod opsin + dehydroretinal = porphyropsin
cone opsin + dehydroretinal = cyanopsin.

Each of the four carbon–carbon double bonds of the side chain can in theory exist in either the *cis* or the *trans* configuration. Only two of the many possible isomers have been found, namely the very stable all-*trans* form and the 11-*cis* form, which is stable only in the dark.

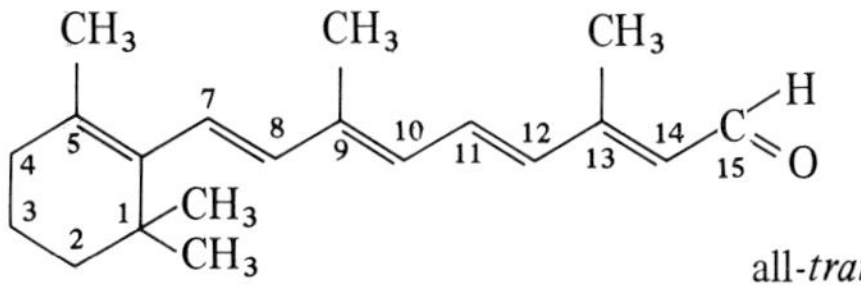

all-*trans*-retinal

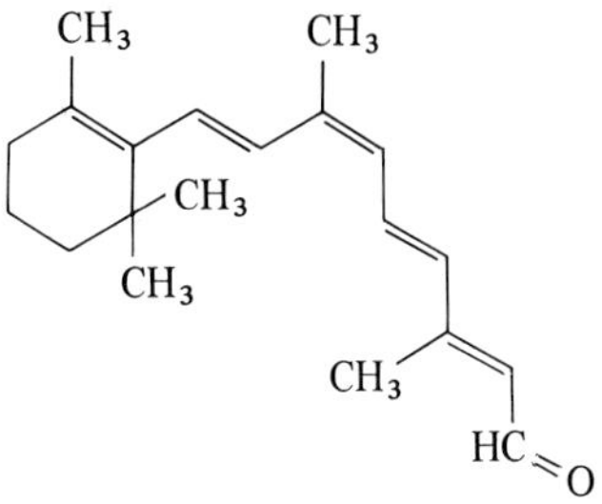

9-*cis*-retinal

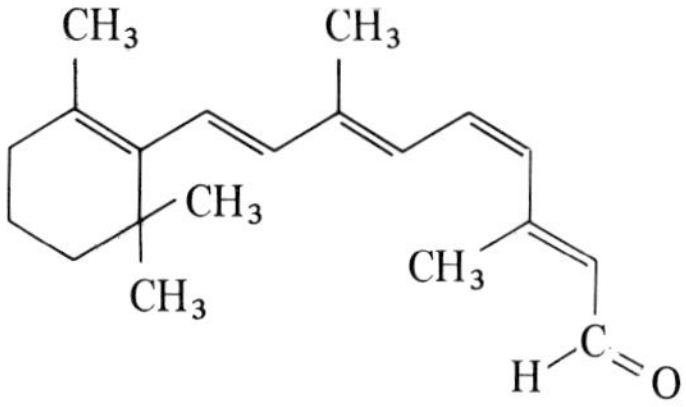

11-*cis*-retinal

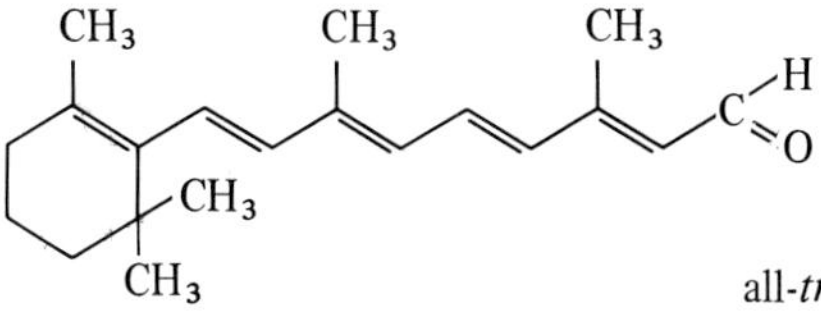

all-*trans*-3-dehydroretinal

Scheme 8.4

The 11-*cis* form is the chromophore of the visual pigments. The synthetic 9-*cis* form similarly combines with opsins to give the so-called isopigments. None of these isopigments has so far been found in the retina.

When visual pigment (opsin + 11-*cis*-retinal or 11-*cis*-dehydroretinal) is exposed to light the chromophore isomerizes totally to the *trans* form. This step, also called 'bleaching', is the sole light reaction. All subsequent reactions involved in the visual process take place in the dark. As all-*trans*-retinal cannot couple with opsin, it is split off from the opsin during the bleaching process. All-*trans*-retinal must isomerize back to regenerate visual pigment. In this way a cyclic *cis*–*trans*-isomerization is seen to be an important part of the visual process (scheme 8.5). The greater part of the all-*trans*-retinal is reduced to the alcohol by a nonspecific aldehyde reductase dependent on NADH or NADPH[38]. The binding between opsin and retinal has the character of a Schiff's base; the partner is the ε-amino group of a lysine residue of the opsin[39-41].

Vitamin A is also involved in other physiological processes, namely growth, development, and reproduction[42]. However, the molecular mechanism of these effects is not yet known.

It has been debated whether vitamin A might have coenzyme-like functions, since a deficiency of vitamin A leads to a lowering of oxidative phosphorylation[44] and a lowering of the activity of the following enzymes: gulonolactone oxidase[45], codeine demethylase[46], squalene cyclohydrogenase[47], ATP sulphurylase[48], Δ-5,3-β-hydroxysteroid dehydrogenase[49], and sulphate transferase[50]. In addition, it is supposed that there is an effect on membranes. Cartilage rudiments in organic cultures release into the surrounding medium, in the presence of vitamin A, a lysosomal acid protease, cathepsin D[51]. Isolated lysosomes on treatment with vitamin A likewise release proteolytic enzymes into the surrounding medium[52]. Similar effects have also been found *in vivo*[53-55].

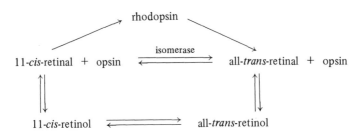

Scheme 8.5

Biosynthesis of retinal

The animal organism cannot itself synthesize retinal and is dependent on the presence of a precursor, β-carotene. The synthesis of carotene is achieved by higher plants, algae, and photosynthesizing bacteria. The biosynthetic route is reasonably well understood and proceeds according to the following brief description[43]. The basic fragment is activated acetate; three of these basic units come together as 3-hydroxymethylglutaryl-CoA. Reductive cleavage of the CoA residue leads to mevalonic acid, from which isopentenyl pyrophosphate is formed after phosphorylation, and elimination of water and carbon dioxide. Four of these isopentenyl pyrophosphate residues condense to a C_{20} terpenyl pyrophosphate, from which the carotene precursor with forty carbon atoms, phytoene, is obtained by head-to-head condensation. Dehydrogenation, cyclization, isomerization, hydration, and hydroxylation reactions then follow and afford the various carotenes as end products. In the mammalian organism β-carotene taken up with the food is cleaved into two molecules of retinal by a cytoplasmically localized β-carotene-15,15'-dioxygenase of the mucosa.

8.5 Collagenaldehyde
Rigidity and insolubility of the connective tissue is caused by strong interfibrillar and intrafibrillar cross-linkages in collagen and elastin. Certain lysine side chains (5α) are oxidized through lysyl oxidase[56] to α-aminoadipate-5-aldehyde (allysin). Subsequent aldol condensation between the neighbouring carbonyl groups thus created[57,58], or Schiff's base formation with other ε-amino groups of lysine[59], leads to cross-linking. Aldehyde functions are probably involved also with intermolecular binding. It is assumed that formation of Schiff's bases takes place and that these bases eventually become stable towards acids. The following findings favour this hypothesis:
(a) thiosemicarbazide blocks free carbonyl groups and thereby prevents the gradually developing decrease in solubility of initially soluble collagen fractions *in vitro*[60];
(b) initially water-soluble collagen fractions, which become water-insoluble on ageing, dissolve in acid media. This may readily be explained through hydrolytic cleavage of Schiff's bases;
(c) if the aged collagen fractions [as described in (b)] are treated with sodium borohydride, they lose the ability to dissolve in acids. This is compatible with the supposition that Schiff's bases are reduced by sodium borohydride to secondary amines, which are, as is well known, stable to acids[61].

8.6 Aldehydes in plants and fruit
8.6.1 Aldehydes from cuticular leaf wax
The first reports on the occurrence of long chain aldehydes in plant waxes were made by Lamberton and Redcliffe[62] in 1960, and later work appears to confirm that these long chain aldehydes are general components of plant waxes. The aldehyde content of surface lipids depends on the species examined; the various types show different patterns in the distribution of aldehydes, as can be seen from table 8.1. Starrat and Harris[63] believe that the determination of distribution patterns of aldehydes will become important for taxonomic studies.

Table 8.1. Aldehydes in surface waxes of leaves and fruits.

Source	Aldehydes (% of total lipids)	Distribution of aldehydes (%)									Reference
		C_{24}	C_{25}	C_{26}	C_{27}	C_{28}	C_{29}	C_{30}	C_{31}	C_{32}	
Pea (*Pisum sativum*)	5·0	1·4		55·7	1·0	40·9		1·0			71
Dogbane (*Apocynum androsaemifolium* L.)	2·5			2·5		15·5	7·2	74·8			63
Indian hemp (*Apocynum cannabinum* L.)	1·1	4·2		5·6		18·1	6·6	68·5			63
Cypress spurge (*Euphorbia cyparissias* L.)	2·8		4·1	8·5		75·6		11·8			63
Spurge (*Euphorbia esula* L.)	0·5			11·3		73·5		15·2			63
Toadflax (*Linaria vulgaris* L.)	3·3	6·1	1·7	20·8	6·8	8·2	11·6	22·5		22·3	63
Purple loosestrife (*Lythrum salicaria* L.)	0·2	2·3	3·9	16·4	8·6	21·8	12·0	32·9	2·1		63
Fat hen (*Chenopodium album*)	11·6	2·9	0·6	13·6	1·2	46·0	0·3	14·6		9·3	75
Ryegrass (*Lolium perenne* L.)	9·1		1·6	55·7	1·2	22·3	1·1	15·8		1·7	75
Grape	12·4		2·8	41·7	2·5	21·8	1·1	7·5	0·5	2·8	73
Cranberry	14·3			1·1	1·9	10·2	5·2	76·2	0·8	3·9	74

Little is known about the biosynthesis of these long chain aldehydes. Kolattukudy[64,65] found activity in aldehydes, when [^{14}C]acetate and ^{14}C-labelled fatty acids were supplied to Brassicaceae. Kolattukudy also found that extracts of young broccoli leaves (*Brassica oleracea*) catalysed the reduction of acyl-CoA to the aldehyde and then to the alcohol[66], and from the same extracts an enzyme was isolated which catalyses the reduction of acyl-CoA[67]. Enzymes with the same catalytic properties were also isolated from *Euglena gracilis*[68] and *Clostridium butyricum*[69]. α-Oxidation of long chain fatty acids[70] gives odd-numbered aldehydes and should therefore be of significance regarding the origin of these aldehydes, which occur only in small amounts. Since the distribution of chain lengths is the same for aldehydes and alcohols, it is possible that these aldehydes may only be intermediate products in the biosynthesis of long chain alcohols[71].

8.6.2 2-*trans*-Hexenal

2-*trans*-Hexenal (2TH) is a main representative of naturally occurring aldehydes. The wide distribution of this aldehyde in the higher plants has earned it the name 'leaf aldehyde'[77]. 2TH has been found in macerated leaves[78-80], in mown grass[81-83], and in homogenates of strawberries[84], grapes[85], apples[86], raspberries[87], and many other fruits[85].

The macerated leaves of *Gingko biloba*[78,79], *Albizzia julibrissin*[88], and *Ailanthus glandulosa*[80] show the greatest capacity for formation of 2TH. The leaves of *Gingko biloba* are capable of forming 2TH during the whole year, but those of *Ailanthus glandulosa* and *Rhododendron nudiflorum* L. develop this capacity only in late summer[78]. Leaves of *Ligustrum vulgare* L. (privet) and of *Quercus alba* L. (white oak)[78] form no 2TH after maceration. All observations up to the present suggest the involvement of one or more enzymes in the formation of 2TH. Nye and Spoehr[80] found that boiling the leaves completely stops the formation of 2TH. In addition, methanol, trichloroacetic acid, and sodium sulphite inhibit the formation of 2TH, but potassium cyanide does not[85]. A definite dependence on pH was described by Major *et al.*[88]. A pH optimum of 4·0 was found with homogenates of *Ailanthus glandulosa*, and 4·4 with *Gingko biloba*. Oxygen is an essential reactant[80]; when the maceration of the leaves and the preparation were carried out under nitrogen no 2TH was observed[84,85].

There has been much guesswork about the compound which precedes 2TH. Originally 3-hexen-1-ol, a compound present in many plants, was considered as the precursor, a view suggested by the ease with which 3-hexen-1-ol is oxidized to 2TH.

Nye and Spoehr have recognized unsaturated fatty acids, fats, and fatty acid esters as possible precursors[80]. These workers have compared yields of 2TH from macerated leaves before, and after, addition of oleic acid. The added oleic acid appreciably enhanced the yield of 2TH, provided that the macerated leaves were able to form 2TH even in the absence of added oleic acid. Major and Thomas[79] showed that 2TH from *Gingko biloba* extracts was formed from uniformly ^{14}C-labelled linolenic acid, though the 2TH is formed from linolenic acid under these experimental conditions only if the enzymatic activities of the leaf homogenate remain intact. This result excludes, in the opinion of these authors[79], the possibility of an autoxidation. The absolute requirement of oxygen, and a positive peroxide test, led to the supposition that lipoxidase is responsible for this reaction. More detailed investigations showed, however, that a significant involvement of lipoxidase may be excluded, since it has a pH optimum between 6·5 and 7·0[89]. Extracts from *Gingko* leaves show, as already mentioned, maximal 2TH synthesis at pH 4·4, and only very slight activity at pH > 6·5.

Schildknecht and Rauch[90] investigated whether 2TH synthesis was caused by the actual destruction of the leaf structure, by enclosing a small tree of the species *Robinia pseudoacacia* in a polyethylene bag. Twelve hours later they could show the presence of 2TH in the air enclosed in the bag. However, this result is equivocal since Drawert et al.[85] assumed that damage to a few leaves could not be excluded with certainty. Thus the biosynthesis of 2TH in intact leaves still remains rather problematical.

2TH is an active fungicide and is toxic also to other classes of the Protista. Damaged leaves of several species, in particular of *Gingko biloba*, develop considerable resistance towards fungal attack and this has been connected with the capacity of damaged leaves to produce 2TH[78]. Thus it seems possible that the biosynthesis of 2TH represents some kind of biological defence for certain plants against fungal attack after damage to their leaves.

8.6.3 Aldehydes in fruit

Aldehydes are widely distributed in Nature. The aroma and the taste of numerous fruits are determined, or at least substantially influenced, by these compounds (see section 8.8). Thus all kinds of aldehydes, but especially 2TH, have been found in apples, pears[91,92], grapes[93], raspberries[94,95], strawberries[84], tomatoes[96,97], peas[98,99], cranberries[100], peaches, cherries, sour cherry[87], and in the essential oils of orange peel[101], grapefruit[102], and hops[103].

Citral is mostly found together with citronellal and other aldehydes in all citrus fruit[104-108] and is the factor responsible for the natural taste of lemon. It is also found in many essential oils, for example lemon grass oil, eucalyptus oil, oil of verbena, ginger oil, rose oil, and many others[109-118].

Excellent surveys on fragrant compounds from fruit have been published by Nursten and Williams[119], and by Gierschner and Baumann[120]. A very high aldehyde content (13–18 mg/kg) has been found in strawberries and pears[87]. Table 8.2 provides a survey of aldehydes present in various fruits; a study of this table reveals that 2TH has an improbably wide distribution, and there are many indications that this aldehyde arises during the processing of the fruit (cf section 8.6.2).

Table 8.2. Carbonyl compounds in various fruits. Unless otherwise stated the alkenals have the 2-*trans* configuration.

Pineapple[119]	*Banana*[76,119]	*Blackcurrant*[119]	*Grape*[119]
formaldehyde	acetaldehyde	acetaldehyde	acetaldehyde
acetaldehyde	1-pentanal	butanal	butanal
furfural	2-hexenal	pentanal	hexanal
	C_{24}, C_{26}, C_{28},	hexanal	2-hexenal
	C_{29}, C_{30}, C_{31},	2-hexenal	benzaldehyde
	C_{32} aldehydes	benzaldehyde	
Apple[72,119]	*Pear*[119]	*Strawberry*[119]	*Raspberry*[119]
formaldehyde	acetaldehyde	acetaldehyde	acetaldehyde
acetaldehyde	propanal	propanal	propanal
propanal	2-hexenal	2-propenal	2-propenal
1-butanal	*Peach*[119,75]	2-butenal	2-methylpropenal
pentanal	acetaldehyde	2-pentenal	2-pentenal
hexanal	benzaldehyde	hexanal	2-hexenal
2-hexenal	furfural	3-*cis*-hexenal	3-*cis*-hexenal
furfural	C_{24}, C_{26},	heptanal	benzaldehyde
C_{24-30} aldehydes	C_{30} aldehydes	benzaldehyde	furfural
		furfural	methylfurfural
		methylfurfural	

Grapefruit[119]	*Lemon*[119]	*Orange*[76,101,119,121]	
acetaldehyde	heptanal		
citral	octanal	acetaldehyde	citral
C_{7-11} aldehydes	nonanal	pentanal	neral
	decanal	hexanal	geranial
Lime[119]	undecanal	2-hexenal	dodecanal
octanal	dodecanal	heptanal	α,β-substituted
nonanal	C_{13-17} aldehydes	octanal	acroleins
citral	citral	octenal	C_{24}, C_{26}, C_{28},
dodecanal	neral	nonanal	C_{30} aldehydes
furfural	geranial	decanal	
	citronellal	undecanal	

8.7 Aldehydes as substances transmitting information

The transfer of information between living creatures can take place in various ways, and sometimes compounds of low molecular weight are involved. Law and Regnier[122] termed such substances 'semiochemicals' (Greek: σημε'ων; semeion = signal, or sign) these being substances which are secreted by an individual of a species and which contain specific information for another individual.

Semiochemical communication may take place between individuals of different species (interspecies) or of the same species (intraspecies). Substances active intraspecies are also called 'pheromones'. However, an exact differentiation between intraspecies and interspecies action is not possible.

Table 8.3. Aldehydes as semiochemicals. Unless otherwise stated, unsaturated aldehydes have the *2-trans* configuration.

Aldehyde	Occurrence	Reference
propanal	Scutteleridae [a] (*Crysocoris stolli*)	134
	tree bug (*Libyaspis angolensis*)	133
n-butanal	tree bug (*Libyaspis angolensis*)	133
	squash bug (*Amorbus rhombifer*)	125
pentanal	Cydnidae [a] (*Scaptocoris divergens*)	
hexanal	Scutteleridae [a] (*Crysocoris stolli*)	134
	tree bug (*Mictus profana* Fabricius)	145
	squash bug (*Amorbus rubiginosus* Guerin)	145
	weaver ant (*Oecophylla longinoda*)	200
	coreid bug (*Amblypelta nitida*)	201
tetradecanal	Scutteleridae [a] (*Crysocoris stolli*)	134
acrolein	Scutteleridae [a] (*Crysocoris stolli*)	134
	shield bug (*Nezara viridula*)	134
	Cydnidae [a] (*Scaptocoris divergens*)	145
crotonaldehyde	Scutteleridae [a] (*Crysocoris stolli*)	134
	shield bug (*Nezara viridula*)	134
	Cydnidae [a] (*Scaptocoris divergens*)	145
pentenal	Scutteleridae [a] (*Crysocoris stolli*)	134
hexenal	Scutteleridae [a] (*Crysocoris stolli*)	134
	shield bug (*Nezara viridula*)	134
	tree bug (*Rhoecocoris sulciventris*)	145
	tree bug (*Musgraveis sulciventris*)	134
	tree bug (*Poecilometis strigatus*)	134
	shield bug (*Brochymena quadripustulata*)	134
	sloe bug (*Dolycoris baccarum*)	134, 135
	tree bug (*Eurygaster* sp.)	134
	leather bug (*Acanthacephala femorata*)	136, 137
	leather bug (*Acanthacephala granulosa*)	137
	leather bug (*Acanthacephala declivis*)	137
	cockroach (*Eurycotis decipiens*)	138
	cockroach (*Eurycotis floridana*)	138
	cockroach (*Eurycotis biolleyi*)	138
	tree bug (*Libyaspis angolensis*)	132
	cockroach (*Cutilia soror*)	145
	cockroach (*Platyzosteria novae seelandiae*)	145
heptenal	shield bug (*Nezara viridula*)	134
	rice stink bug (*Oebalus pugnax*)	134, 139
	shield bug (*Euchistus servus*)	134
	Cydnidae [a] (*Scaptocoris divergens*)	145
octenal	sloe bug (*Dolycoris baccarum*)	135
	Scutteleridae [a] (*Crysocoris stolli*)	134
	shield bug (*Nezara viridula*)	134
	tree bug (*Musgraveis sulciventris*)	134
	tree bug (*Rhoecocoris sulciventris*)	146

Table 8.3 (continued)

Aldehyde	Occurrence	Reference
octenal	tree bug (*Poecilometis strigatus*)	134, 145
	tree bug (*Eurygaster* sp.)	134
	tree bug (*Palomena viridissima*)	134
	Cydnidae [a] (*Scaptocoris divergens*)	145
decenal	tree bug (*Libyaspis angolensis*)	133
	Scutteleridae [a] (*Crysocoris stolli*)	134
	shield bug (*Nezara viridula*)	134, 145
	tree bug (*Musgraveis sulciventris*)	134
	spiny orange bug (*Biprorulus bibax*)	134, 146
	sloe bug (*Dolycoris baccarum*)	134, 135
	tree bug (*Palomena viridissima*)	134
trans-dodecenal	swallowtail butterfly (papilinoid caterpillar)	144
salicylaldehyde	carabid beetle (*Calosoma prominens* Lec.)	142
	musk beetle (*Aromia moschata* L.)	143
	red poplar leaf beetle (*Melasoma populi* L.)	144
	brassy willow beetle (*Phyllodecta vitellinae*)	144
	poplar leaf beetle (*Plagiodera* sp.)	144
2-butyl-2-octenal	weaver ant (*Oecophylla longinoda*)	200
	coreid bug (*Amblypelta nitida*)	201
3-undecanone	weaver ant (*Oecophylla longinoda*)	200
Sex pheromones[122]		
n-undecenal	wax moth (*Galleria mellonella*)	
benzaldehyde	smoky wainscot moth (*Leucania impura*)	
cis-3,3-dimethyl $\Delta^{1,\beta}$-cyclohexan-acetaldehyde	boll weevil (*Anthomonus grandis*)	
trans-3,3-dimethyl $\Delta^{1,\beta}$-cyclohexan-acetaldehyde	boll weevil (*Anthomonus grandis*)	
Alarm pheromones[122]		
citral	ants (Formicinae)	
	bees, wasps (Hymenoptera)	
citronellal	ants (Formicinae)	
hexenal	stinging ants (Myrmicinae)	
Recruiting pheromones[122]		
geranial	honey bee (*Apis mellifera*)	
neral	honey bee (*Apis mellifera*)	

[a] No common name

Within the scope of this book these substances are of interest insofar as an appreciable number of aldehydes are found among them. Semiochemicals serve as products of evolution in maintaining and improving the species. The interspecies agents act most importantly to repel enemies chemically, and the pheromones act as effective attractants of the sexual partner, for the marking out of territories, as alarm substances for collective defence, for improving the efficiency of food gathering ('recruiting'), etc. The aldehyde components of semiochemicals are summarized in table 8.3, which is arranged according to chemical type as well as source and mode of action.

8.7.1 Biogenesis of semiochemicals

Very little is known about the biosynthesis of these compounds. Gordon et al.[123] showed that [^{14}C]acetate is incorporated into aldehydes. *Melasoma populi* probably utilizes populin and salicin as starting materials for the synthesis of salicylaldehyde[124]. Waterhouse et al.[125] assume that *Amorbus rubiginosus* may be able to concentrate aldehydes taken in as nutriment. Citral and citronellal are built up from acetate by the ant *Acanthomyops* analogously to the biosynthesis of terpenes[126]. What is remarkable is that the individual is itself not harmed by the production of what are often highly toxic substances.

8.7.2 Recognition of pheromones

It has been shown, for instance with bombycol, that quite small amounts of substance suffice to call forth a positive alteration in behaviour[127]. With moths (*Bombyx mori*) 10^4 molecules cm^{-3} at a wind velocity of 60 cm s^{-1} produce a positive change in behaviour in 50% of animals[128]. It was concluded that most probably only a single molecule need react with the receptor in order to trigger an impulse[129]. But by what means does an organ, a cell or part of a cell perceive something as small as a molecule? Various hypotheses have been put forward and are all based on known systems. For example, the hypothesis of Amoore[130], which bears a strong similarity to the lock-and-key model for enzyme action; or that of Riddiford[131], which has as its premise the oestrogen–protein binding hypothesis, or the repressor–inducer model.

In general the perception of a pheromone is considered to be an olfactory process. The active sites of various receptor organs of insects were described by Schneider[132], but we have hardly any information on the molecular processes which must occur there. However, as the next section suggests, sensory receptors for carbonyl compounds are very well developed and widely occurring.

8.8 Carbonyl compounds as vehicles of taste and odour

Aliphatic carbonyl compounds represent the most important group of flavouring compounds in our foodstuffs. One finds them in all flavour extracts. They are either entirely, or in large measure, responsible for nearly all known flavours and determine, even when present in small amounts, the taste and odour of our foodstuffs, and beverages such as tea and coffee. As well as contributing to the fine fragrance and good taste of our foods, they can also be responsible for nauseous flavours, for example of rancid fat or oil, and of putrid meat.

Straight chain saturated aldehydes with chain lengths of two to five carbon atoms generally produce an odour reminiscent of green leaves or of fruit. Aldehydes with chain lengths of six to eleven carbon atoms possess a bitter flavour, or else a flavour similar to orange peel. A pleasant fruity flavour is ascribed to short chain singly unsaturated aldehydes. These compounds are unique insofar as their flavours are altered drastically with small alterations of structure. Meijboom[147] synthesized eleven different isomers of nonenal and compared their flavours and their threshold concentrations. 2-*trans*-Nonenal in paraffin oil has an odour of fresh linen. 3-*cis*-Nonenal exhibits a flavour similar to that of green cucumber, and the 8-nonenal has a flavour resembling melon. These examples illustrate the specificity of receptor–aldehyde interactions.

The flavour is altered even when going from the *cis* to the *trans* configuration[148]. For example, 2-*trans*-4-*cis*-decadienal has a different flavour from the all-*trans* isomer. Not only does the flavour change with such small alterations in the molecule, but also the concentration at which a distinct flavour is still just detectable changes. The flavour threshold of straight chain aldehydes is dependent on the chain length and decreases with increasing number of carbon atoms. Of course, the medium in which the aldehyde is examined also markedly influences the threshold concentration[149].

In the flavour concentrates nearly the whole range of saturated, branched-chain, and unsaturated aldehydes has been found. Day[150] subdivides the carbonyl derivatives of flavour extracts from foodstuffs into seven classes: alkanals, alk-2,4-dienals, alk-2-enals, alk-1-en-3-ones, α-dicarbonyl compounds, β-dicarbonyl compounds, and alkan-2-ones.

8.8.1 Origin of flavour-producing carbonyl compounds

Egli specified four reactions which could lead to the formation of these flavour-producing carbonyl compounds: Strecker degradation, Maillard reaction, enzymatic oxidation of lipids, and autoxidation of lipids. Since many of our foodstuffs are exposed to fermentation processes during their production, the enzymatic activities of the microorganisms involved must also be taken into consideration. The first two paths mentioned lead to carbonyl compounds with a maximum of five carbon atoms. Since most flavouring compounds are, however, of longer chain length, it appears

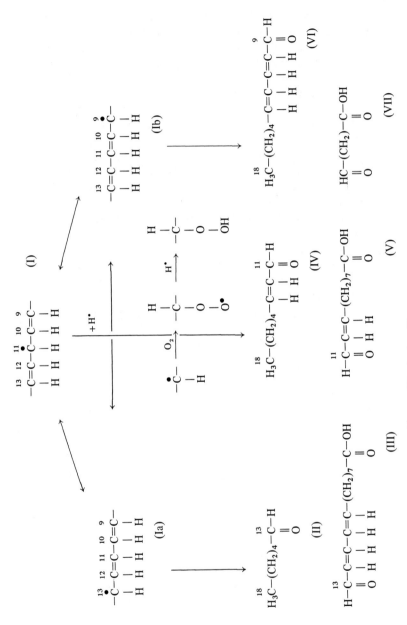

Scheme 8.6. Mechanism of the autoxidation of linolenic acid.

probable that they derive from lipids either through enzymatic catalysis or by autoxidation. It will be very difficult to decide between these paths, since both proceed via hydroperoxides and free radicals, are catalysed by copper- or iron-containing proteins, and mostly they have the same end product.

The free radical mechanism of autoxidation formulated by Farmer and Sutton[151] is now generally accepted. Bolland[152] and Bateman[153] provide very good surveys of these studies. For example, scheme 8.6 shows the reaction path for linolenic acid. The initial reaction is the formation of a methylene radical (I), which is mesomerically stabilized by contributions of the forms (Ia) and (Ib). Each of these forms may produce peroxy radicals with oxygen. These can then abstract hydrogen atoms (radicals) from a molecule of linolenic acid, giving the hydroperoxide and a new methylene radical (I). In this way three isomeric linolenic acid mono-hydroperoxides are formed, and these can give rise, through chain cleavage at the hydroperoxy groups at C-13, C-11, and C-9, to saturated as well as singly and doubly conjugated unsaturated aldehydes (II–VII). Some aldehydes, aldehyde acids, and aldehyde esters have been isolated from autoxidation of linolenic acid or its esters.

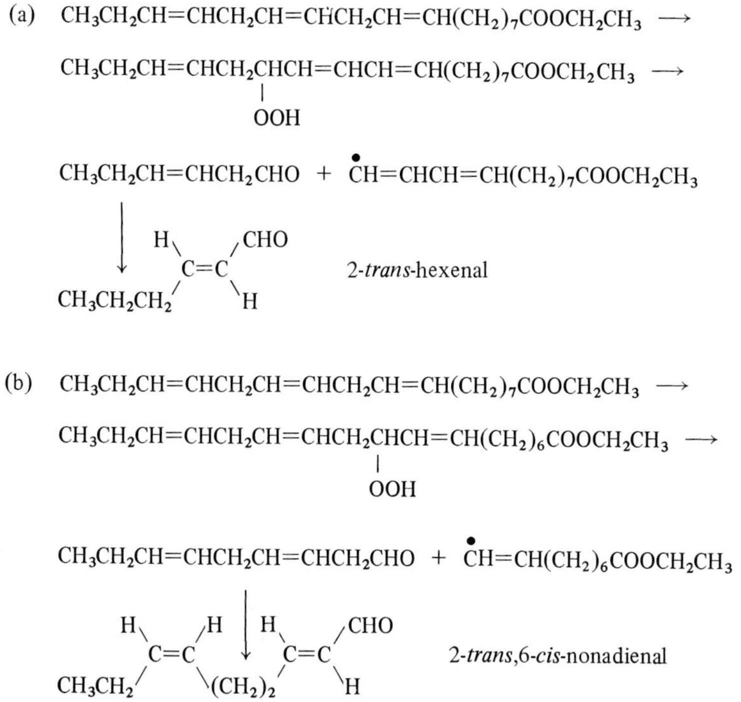

(a) $CH_3CH_2CH=CHCH_2CH=CHCH_2CH=CH(CH_2)_7COOCH_2CH_3 \longrightarrow$

$CH_3CH_2CH=CHCH_2CHCH=CHCH=CH(CH_2)_7COOCH_2CH_3 \longrightarrow$
$\qquad\qquad\qquad\qquad\ \ |$
$\qquad\qquad\qquad\quad\ \ OOH$

$CH_3CH_2CH=CHCH_2CHO + \overset{\bullet}{C}H=CHCH=CH(CH_2)_7COOCH_2CH_3$

$\downarrow\quad \underset{CH_3CH_2CH_2}{H}\diagdown C=C \diagup^{CHO}_{\diagdown H}\qquad$ 2-*trans*-hexenal

(b) $CH_3CH_2CH=CHCH_2CH=CHCH_2CH=CH(CH_2)_7COOCH_2CH_3 \longrightarrow$

$CH_3CH_2CH=CHCH_2CH=CHCH_2CHCH=CH(CH_2)_6COOCH_2CH_3 \longrightarrow$
$\qquad\qquad\qquad\qquad\qquad\qquad\ \ |$
$\qquad\qquad\qquad\qquad\qquad\quad\ \ OOH$

$CH_3CH_2CH=CHCH_2CH=CHCH_2CHO + \overset{\bullet}{C}H=CH(CH_2)_6COOCH_2CH_3$

$\underset{CH_3CH_2}{H}\diagdown C=C \diagup^{H}_{\diagdown (CH_2)_2} \downarrow \underset{}{H}\diagdown C=C \diagup^{CHO}_{\diagdown H}\qquad$ 2-*trans*,6-*cis*-nonadienal

Scheme 8.7. (a) Mechanism of formation of 2-*trans*-hexenal[159] from ethyl linolenate; (b) mechanism of formation of 2-*trans*,6-*cis*-nonadienal from ethyl linolenate[159].

The formation of 3-nonenal and of 2-nonenal was explained differently by Esterbauer and Schauenstein[154], namely from $\Delta^{10,12}$-linoleic acid 9-monohydroperoxide (Ib) through chain cleavage between C-9 and C-10. A radical $CH_3(CH_2)_6CH=CHCH=CH^\bullet$ is thereby formed, which with a hydroxyl radical gives 3-nonenal via an unstable alcohol $-CH=CHOH$. Isomerization can also give 2-nonenal.

Further mechanisms for the formation of 2-hexenal and 2,6-nonadienal are presented in scheme 8.7. From the mechanism given in scheme 8.6 unsaturated aldehydes are to be expected as main products; in fact, however, saturated aldehydes predominate. This is so because the initially formed unsaturated aldehydes undergo further oxidation[150].

Table 8.4 provides a survey of autoxidation products of oleic acid, linoleic acid, linolenic acid, and arachidonic acid. When foodstuffs take on a rancid odour, this is due to autoxidation of the unsaturated lipids contained in them. The same mechanism, incidentally, is involved in the drying of oil paints and the detonation of a plastic explosive[194].

Table 8.4. Aldehydic autoxidation products of unsaturated fatty acids (according to Badings[164]).

Starting product	End product	Starting product	End product
linoleic acid	pentanal	oleic acid	heptanal
	hexanal		octanal
	heptanal		nonanal
	octanal		decanal
	2-*trans*-heptenal		2-*trans*-decenal
	2-*trans*-octenal		2-*trans*-undecenal
	2-*cis*-octenal	linolenic acid	acetaldehyde
	2-*trans*-nonenal		propanal
	2-*cis*-decenal		hexanal
	3-*trans*-nonenal		2-*trans*-crotonaldehyde
	3-*cis*-nonenal		2-*trans*-pentenal
	2,4-*trans*-nonadienal		2-*cis*-pentenal
	2,4-decadienal		2-*trans*-hexenal
	2-*trans*,4-*cis*-decadienal		2-*trans*-heptenal
arachidonic acid	pentanal		3-*trans*-hexenal
	hexanal		3-*cis*-hexenal
	2-*trans*-heptenal		2-*trans*,4-*trans*-heptadienal
	2-*trans*-octenal		2-*trans*,4-*cis*-heptadienal
	2-*cis*-octenal		2-*cis*,5-*cis*-octadienal
	3-*cis*-nonenal		2-*trans*,6-*cis*-nonadienal
	4-*cis*-decenal		2-*trans*,4-*cis*,7-*cis*-decatrienal
	2-*trans*,6-*cis*-dodecadienal		2-*trans*,4-*trans*,7-*cis*-decatrienal
	2,4,7-tridecatrienal		
	2-*trans*,4-*trans*-decadienal		
	2-*trans*,4-*cis*-decadienal		

In general terms, a characteristic taste or odour is caused not by the presence of a single substance, but of several substances. Only in precisely determined concentrations, often below a critical concentration, is a characteristic bouquet developed (see, for example, the investigations of Boyt et al.[155] on the flavour of chocolate).

8.8.2 Aldehydes associated with specific flavours

Nevertheless there are a few aromas which are in the main connected with specific aldehydes. The malt-like flavour of milk may be ascribed to the presence of 3-methylbutanal and 2-methylbutanal. Neither of these aldehydes is an autoxidation product of unsaturated fatty acids in milk, but they arise from the appropriate amino acids[156] through transaminase or decarboxylase activity of *Streptococcus lactis* var. *maltigenes*. 4-*cis*-Heptanal, and 2-*trans*,6-*cis*- and 2-*trans*,6-*trans*-nonadienals have been found in fat[157,158]. The first-mentioned compound exhibits a creamy tallowy butterscotch-like flavour, and the two latter compounds a green cucumber-like and tallowy odour. The insipid taste that beer develops when it is exposed to the sun is ascribed to 2-nonenal obtained through a combination of lipoxidase-catalysed and autoxidative processes[159]. 2-*trans*,6-*cis*-Nonadienal is the most important compound possessing a cucumber flavour[160]. It is formed enzymatically when pumpkin tissue decomposes[161]. 2-*trans*-Hexenal has been found together with 2-nonenal and 2-*trans*,6-*cis*-nonadienal in stale (flat) beer[159,162]. Although the creamy flavour of butter is ascribed to 4-*cis*-heptenal, its actual presence has not been demonstrated in butter[163]. Detailed investigations on the flavouring compounds of butter have been carried out by Badings[168]. 2-*trans*,4-*trans*-Decadienal causes the oily 'off-flavour' in stale milk[165], the 'deep-fried flavour' of potato chips and other oil-fried foods, and the 'canteen smell' of spoilt vegetables.

The odour and taste of foodstuffs may be influenced by carbonyl compounds that arise during the manufacturing process through fermentation or autoxidation. For example, during every stage of bread manufacture carbonyl compounds are formed, but in particular during fermentation of the half-finished product[166]. For this reason, bread aroma disappears when the duration of fermentation is decreased.

Young ('green') wine contains little free aldehyde, and the greater part of the carbonyl compounds is present in combined form. With increased ageing the total aldehyde content, and the proportion of bound aldehydes, decreases, but the relative proportion of free aldehydes increases[167]. The content of aromatic and aliphatic aldehydes in brandy (cognac) increases with increasing age[22].

Unsaturated aldehydes also arise through thermal degradation of carbohydrates, amino acids, and fats[169-171]. Such thermal degradative processes are probably responsible for the presence of these aldehydes in boiled, fried, and baked foods. Unsaturated aldehydes have been detected in a large number of foodstuffs, such as potatoes[172-174], potato chips[175], poultry[176,177], meat[178,179], fish[180,181], salad oils[182-184], bread, and bakery products[185-187] (table 8.5).

The aldehyde content of apple juice[193] and the juices of citrus fruits[188] has been widely examined. These aldehydes are partly formed biogenically in the fruit and are partly produced by enzymatic as well as nonenzymatic reactions during the processing of the fruit. This is especially shown by the example of hexenal, the content of which in apple juice can serve as a measure of the quality[193].

Table 8.5. Aldehydes found in raw or boiled potatoes[174-176], in rancid sunflower oil[184], and in tea[189-192] as flavour components. The enals have the 2-*trans* configuration.

Potato	Sunflower oil	Tea
acetaldehyde	pentanal	acetaldehyde
butanal	hexanal	propanal
isobutanal	heptanal	butanal
pentanal	octanal	isobutanal
isopentanal	nonanal	2-methylbutanal
hexanal	hexenal	3-methylbutanal
heptanal	heptenal	pentanal
octanal	octenal	hexanal
acrolein	nonenal	heptanal
crotonaldehyde	decenal	acrolein
	undecenal	hexenal
	2-*trans*,4-*cis*-nonadienal	octenal
	2-*trans*,4-*trans*-nonadienal	benzaldehyde
	2-*trans*,4-*cis*-decadienal	phenylacetaldehyde
	2-*trans*,4-*trans*-decadienal	

8.9 Participation of aldehydes in bioluminescence

It was recognized quite early that an enzyme-catalysed reaction formed the basis of bioluminescence. The substrate, which had not been characterized at that time, was called luciferin, and the enzyme, luciferase. In addition to luciferase, FMN, oxygen, and an aldehyde were recognized as essential factors for bacterial luminescence[195,196]. Eley and Cornier[196] showed that the quantum yield of luciferase preparations obtained from *Photobacterium fischeri* or *P. phosphoreum* was dependent on the presence of the long chain aldehyde, dodecenal. An increase of up to 10^4 times in the quantum yield was observed after the addition of this aldehyde. Shimomura *et al.*[197] examined the molecular mechanism of luminescence with the luciferin-luciferase system obtained from *Achromobacter fischeri* and were able to assign the role of luciferin to the aldehyde. They assumed that the aldehyde is oxidized to the acid, as in the following reaction scheme:

$$NADH_2 + FMN \xrightarrow{FMN\ reductase} NAD + FMNH_2$$

$$FMNH_2 + RCHO + O_2 \xrightarrow{luciferase} RCOOH + FMN + H_2O + light$$

The evidence presented for the direct participation of the aldehyde, and for the oxidation of the latter to acid, was as follows: (a) aldehydes with chain lengths of nine to fourteen carbon atoms give a constant quantum yield of 0.154 ± 0.01 Einstein/mol of aldehyde; (b) the acids corresponding to the aldehydes added to the reaction mixture could be detected after acidification and ether extraction; (c) the energy released in the oxidation of the aldehyde (approximately 70 kcal mol^{-1}) exceeds the energy requirement of the light emission.

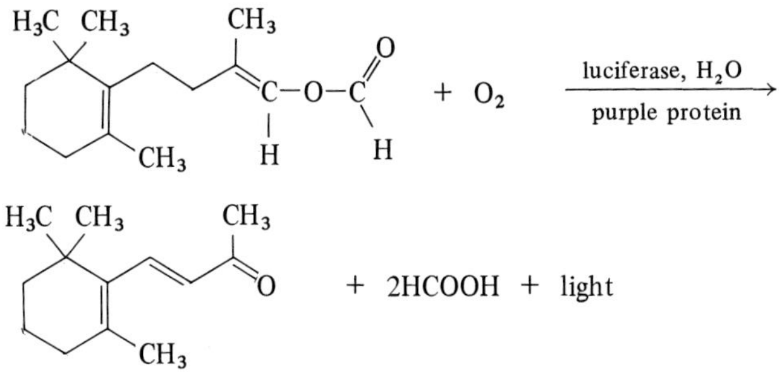

Scheme 8.8. Molecular basis of the bioluminescence of *Latia neritoides*.

McCapra and Hysert[198] observed a linear relation between the quantum yield and the formation of acid from decanal with a luciferase preparation from *P. phosphoreum*. These authors came to the same conclusion as did Eley and Cornier[196], namely that an aldehyde is the luciferin of bacterial luminescence. An aldehyde derivative is also involved in the bioluminescent process of the snail *Latia neritoides* (scheme 8.8). In the light-producing reaction this derivative is oxidized to a ketone and formic acid, with the emission of light[199].

It must be emphasized, of course, that besides the above-mentioned type of bioluminescence, there are other types with nonaldehydic luciferins.

References
1 Kosh, J. W., Appelt, G. D., 1972, *J. Pharm. Sci.*, **61**, 1963.
2 Bertharion, G., Gibert, G., Reynier, M., Laborit, H., 1963, *Agressologie*, **4**, 421.
3 Laborit, H., Wermuth, C., Weber, B., Brue, F., Leterrier, F., Delcambre, J. P., Bertharion, G., Baron, C., 1963, *Agressologie*, **4**, 321.
4 Weber, B., Corbel, M., 1963, *Agressologie*, **4**, 431.
5 Weber, B., 1963, *Agressologie*, **4**, 437.
6 Danon-Boileau, H., Lab, P., Lavitry, S., Lévy, E., Rousseau, A., Laborit, H., 1964, *Agressologie*, **5**, 191.
7 Pougetoux, J., 1969, *Agressologie*, **10**, 93.
8 Jouvet, M., 1969, *Science*, **163**, 32.

9. Fuxe, K., 1965, *Zellforsch. Mikrosk. Anat.*, **65**, 573.
10. Jouvet, M., Vimout, P., Gelorme, J. F., 1965, *J. Pharmacol. Exp. Ther.*, **159**, 1595
11. Sabelli, H. C., Giardina, W. J., Alivisatos, S. G. A., Seth, B. K., Unger, F., 1969, *Nature (London)*, **223**, 73.
12. Sabelli, H. C., Giardina, W. J., 1970, *Biol. Psychiatry*, **2**, 119.
13. Pfeiffer, C. C., Beck, R. A., Goldstein, L., Neiss, E. S., 1966, in *Recent Advances in Biological Psychiatry*, Volume 9, p. 241, Ed. J. Wortis.
14. Johns, D. G., 1967, *J. Clin. Invest.*, **46**, 1492.
15. Bornough, R. C., Erwin, V. G., 1972, *Biochem. Pharmacol.*, **21**, 1457.
16. Feldstein, A., Hoagland, H., Wong, K., Freeman, H., 1964, *J. Stud. Alcohol*, **25**, 218.
17. Turano, C., Giartosio, G., Fasella, P., 1964, *Arch. Biochem. Biophys.*, **104**, 524.
18. Fujioka, M., Snell, E. E., 1965, *J. Biol. Chem.*, **240**, 3050.
19. Fischer, E. H., Krebs, E. G., 1966, *Fed. Proc. Fed. Am. Soc. Exp. Biol.*, **25**, 1511.
20. Fischer, E. H., Forrey, A. W., Hedrick, J. L., Hughes, R. C., Kent, A. B., Krebs, E. G., 1963, *I.U.B., 1st Symp., Rome, 1962*, p.543 (Pergamon Press, Oxford).
21. Hughes, R. C., Jenkins, W. T., Fischer, E. H., 1962, *Proc. Nat. Acad. Sci. USA*, **48**, 1615.
22. Turano, C., Fasella, P., Vecchini, P., Giartosio, A., 1961, *Atti Accad. Naz. Lincei Mem. Cl. Sci. Fis. Mat. Nat. Sez. 1a*, **30**, 532.
23. Schirch, L. G., Mason, M., 1963, *J. Biol. Chem.*, **238**, 1032.
24. Phillips, A. T., Wood, W. A., 1965, *J. Biol. Chem.*, **240**, 4703.
25. Dempsey, W. B., Snell, E. E., 1963, *Biochemistry*, **2**, 1414.
26. Labow, R., Robinson, W. G., 1966, *J. Biol. Chem.*, **241**, 1239.
27. Yamada, H., Adachi, O., Ogata, K., 1965, *Agric. Biol. Chem.*, **29**, 117, 649, 864.
28. Taylor, R. T., Jenkins, W. T., 1966, *J. Biol. Chem.*, **241**, 4391.
29. Jenkins, W. T., Sizer, I. W., 1957, *J. Am. Chem. Soc.*, **79**, 2655.
30. Riva, F., Vecchini, P., Turano, C., Fasella, P., 1964, *Proc. Int. Congr. Biochem. 6th, New York, 1964, Abstr.*, Sect. IV-140, 329.
31. Malakovha, E. A, Torchinsky, Yu. M., 1965, *Dokl. Akad. Nauk SSSR*, **161**, 1224.
32. Katunuma, N., Matsuda, S., Izumi, M., 1962, *Symp. Enzyme Chem. (Tokyo)*, **16**, 70.
33. Buffoni, F., 1968, in *Pyridoxal Catalysis: Enzymes and Model Systems*, Second International Symposium on Chemical and Biological Aspects of Pyridoxal Catalysis, I.U.B., Moscow, 1966, Eds E. E. Snell, A. E. Braunstein, E. S. Severin, Yu. M. Torchinsky (John Wiley, New York)
34. Suzue, R., Haruna, Y., 1970, *J. Vitaminol.*, **16**, 154.
35. Hill, R. E., Rowell, F., Gupta, R. N., Spenser, I. D., 1972, *J. Biol. Chem.*, **247**, 1869.
36. Hill, R. E., Spenser, I. D., 1970, *Science*, **169**, 773.
37. Bernhard, S., 1968, *The Structure and Function of Enzymes* (Benjamin, New York).
38. Fidge, N. H., Goodman, DeW. S., 1968, *J. Biol. Chem.*, **243**, 4372.
39. Collins, F. D., 1953, *Nature (London)*, **171**, 469.
40. Morton, R. A., Pitt, C. A., 1955, *Biochem. J.*, **59**, 128.
41. Bownds, D., 1967, *Nature (London)*, **216**, 1178.
42. Wassermann, R. H., Corradino, R. A., 1971, *Annu. Rev. Biochem.*, **40**, 762.
43. Olson, J., 1964, *J. Lipid Res.*, **5**, 281.
44. Seard, C. R., Vaughn, G., Hove, E. L., 1966, *J. Biol. Chem.*, **241**, 1229.
45. Seshadri Sastra, P., Malathi, P., Subba Rao, K., Ganguly, J., 1962, *Nature (London)*, **193**, 1080.
46. Yonemoto, T., Johnson, B. C., 1967, *Fed. Proc.*, **26**, 635.
47. Subba Rao, K., Olson, R. E., 1967, *Fed. Proc.*, **26**, 635.
48. Subba Rao, K., Seshadri Sastry, P., Ganguly, J., 1963, *Biochem. J.*, **87**, 312.

49 Juneja, H. S., Murthy, S. K., Ganguly, J., 1966, *Biochem. J.*, **99**, 138.
50 Carroll, J., Spencer, B., 1965, *Biochem. J.*, **96**, 79P.
51 Weston, P. D., Parett, A. J., Dingle, J. T., 1969, *Nature*, **222**, 285.
52 Dingle, J. T., 1961, *Biochem. J.*, **79**, 509.
53 Weismann, G., Thomas, L., 1963, *J. Clin. Invest.*, **42**, 661.
54 Dingle, J. T., Sharman, I. M., Moore, T., 1966, *Biochem. J.*, **98**, 476.
55 Anderson, O. R., Pfister, R., Roels, O. A., 1967, *Nature*, **213**, 47.
56 Martin, G. R., 1970, *J. Biol. Chem.*, **245**, 1653.
57 Wornstein, P., Kang, A. H., Piez, K. A., 1967, *Proc. Nat. Acad. Sci. USA*, **55**, 417.
58 Wornstein, P., Piez, K. A., 1966, *Biochemistry*, **5**, 3460.
59 Lent, R., Franzblau, C., 1967, *Biochem. Biophys. Res. Commun.*, **26**, 43.
60 Tanzer, M. L., Monroe, D., Gross, J., 1966, *Biochemistry*, **5**, 1919.
61 Tanzer, M. L., 1967, *Biochim. Biophys. Acta*, **133**, 584.
62 Lamberton, I. A., Redcliffe, A. H., 1960, *Aust. J. Chem.*, **13**, 261.
63 Staratt, A. N., Harris, P., 1971, *Phytochemistry*, **10**, 1855.
64 Kolattukudy, P. E., 1965, *Biochemistry*, **4**, 1844.
65 Kolattukudy, P. E., 1966, *Biochemistry*, **5**, 2265.
66 Kolattukudy, P. E., 1970, *Lipids*, **5**, 259.
67 Kolattukudy, P. E., 1971, *Arch. Biochem. Biophys.*, **142**, 701.
68 Kolattukudy, P. E., 1970, *Biochemistry*, **9**, 1095.
69 Day, J. I. E., Goldfine, H., Hagen, P. O., 1970, *Biochim. Biophys. Acta*, **218**, 179.
70 Martin, R. O., Stumpf, P. K., 1959, *J. Biol. Chem.*, **234**, 2548.
71 Kolattukudy, P. E., 1970, *Lipids*, **5**, 398.
72 Schmid, H. H. O., Bandi, P. C., 1969, *Hoppe Seyler's Z. Physiol. Chem.*, **350**, 462.
73 Radler, F., Horn, D. H. S., 1965, *Aust. J. Chem.*, **18**, 1059.
74 Croteau, R., Fagerson, I. S., 1971, *Phytochemistry*, **10**, 3245.
75 Allebone, J. E., Hamilton, R. J., 1970, *Chem. Phys. Lipids*, **4**, 37.
76 Ferrell, W. J., Drouillard, M., 1970, *Physiol. Chem. Phys.*, **2**, 168.
77 Winter, M., 1962, *Helv. Chim. Acta*, **45**, 2567.
78 Major, R. T., Marchini, P., Boulton, A. J., 1963, *J. Biol. Chem.*, **238**, 1813.
79 Major, R. T., Thomas, M., 1972, *Phytochemistry*, **11**, 611.
80 Nye, W., Spoehr, H. A., 1943, *Arch. Biochem.*, **2**, 23.
81 Snell, E., Wright, L., 1941, *J. Biol. Chem.*, **139**, 675.
82 Pennington, D., Snell, E., Williams, R. J., 1940, *J. Biol. Chem.*, **135**, 213.
83 Schoepfer, H. W., Jung, A. Z., 1938, *Z. Vitaminforsch.*, **7**, 143.
84 Winter, M., Wilhelm, R., 1964, *Helv. Chim. Acta*, **47**, 1215.
85 Drawert, F., Heimann, W., Emberger, R., Tressl, R., 1966, *Justus Liebigs Ann. Chem.*, **694**, 200.
86 Drawert, F., Heimann, W., Emberger, R., Tressl, R., 1965, *Z. Naturforsch.*, **20b**, 497.
87 Pribela, A., Vesatko, J., 1967, *Biologia (Bratislava)*, **22**, 96.
88 Major, R., Collins, O. D., Marchini, P., Schnabl, H. W., 1972, *Phytochemistry*, **11**, 607.
89 Tappel, A. L., 1963, *The Enzymes*, 2nd edn., vol. 5, p.275, Ed. P. T. Boyer, H. Lardy, K. Myrback (Academic Press, New York).
90 Schildknecht, H., Rauch, G., 1961, *Z. Naturforsch.*, **16b**, 422.
91 Kulesza, J., Karwowska, K., 1967, *Riechst. Aromen Körperpflegem.*, **17**, 449.
92 Flath, R. A., Black, D. R., Guadagni, G. D., McFadden, W. H., Schultz, M. T., 1967, *J. Agric. Food Chem.*, **15**, 29.
93 Stevens, K. L., Blomben, J. L., McFadden, W. H., 1967, *J. Agric. Food Chem.*, **15**, 378.
94 Winter, M., Sundt, E., 1962, *Helv. Chim. Acta*, **45**, 195.

95 Mestres, R., Soulie, J., 1965, *Trav. Soc. Pharm. Montpellier*, **25**, 239.
96 Schormüller, J., Grosch, W., 1964, *Z. Lebensm. Unters. Forsch.*, **126**, 38.
97 Pyne, A. W., Wick, E. L., 1965, *J. Food Sci.*, **30**, 192.
98 Grosch, W., 1967, *Z. Lebensm. Unters. Forsch.*, **135**, 75.
99 Whitfield, F. B., Shipton, J., 1966, *J. Food Sci.*, **31**, 328.
100 Anjou, K., Sydow, E. von, 1967, *Acta Chem. Scand.*, **21**, 2076.
101 Moshonas, M. G., Lund, E. D., 1969, *J. Food Sci.*, **34**, 502.
102 Moshonas, M. G., 1971, *J. Agric. Food Chem.*, **19**, 769.
103 Jahnsen, V. J., 1962, *Nature (London)*, **196**, 474.
104 Di Giacomo, A., 1966, *Riechst. Aromen Körperpflegem.*, **16**, 348.
105 Okada, R., Nakamura, T., 1967, *Nippon Shokuhin Kogyo Gakkai Shi*, **14**, 137, 713.
106 D'Amore, G., Calabro, G., 1966, *Ann. Fac. Econ. Commer. Univ. Studi Messina*, **4**, 633.
107 Goretti, G., Annchez, J. L., Liberti, A., 1967, *Riv. Ital. Essenze Profumi Piante Off. Aromi Saponi Cosmet. Aerosol*, **49**, 145.
108 Mehlinetz, A., Minas, T., 1965, *Riechst. Aromen Körperpflegem.*, **15**, 365.
109 Gogol, O. N., 1960, *Tr. Khim. Prir. Soedin.*, **3**, 185.
110 Schratz, E., Schelle, F. J., Quedan, S., 1968, *Sci. Pharm.*, **36**, 13.
111 Schratz, E., Quedan, S., 1965, *Pharmazie*, **20**, 710.
112 Schabort, J., 1967, *J. S. Afr. Chem. Inst.*, **20**, 103.
113 Juvonen, S., 1964, *Planta Med.*, **12**, 488.
114 Attaway, J. A., Pieringer, A. P., Barabas, C. J., 1966, *Phytochemistry*, **5**, 141.
115 Calvarano, M., 1965, *Essenze Deriv. Agrum.*, **35**, 197.
116 Kowl, G. L., Nigam, S. S., 1966, *Riechst. Aromen Körperpflegem.*, **16**, 159.
117 Sharma, M. L., Nigam, M. C., Handa, K. L., Rao, P. R., 1966, *Indian Oil Soc. J.*, **31**, 303.
118 Calvarano, M., 1966, *Essenze Deriv. Agrum.*, **36**, 100.
119 Nursten, H. E., Williams, A. A., 1967, *Chem. Ind. (London)*, **1967**, 486.
120 Gierschner, K., Baumann, G., 1968, *Riechst. Aromen Körperpflegem.*, **18**, 134.
121 Moshonas, M. G., Lund, E. D., 1969, *J. Agric. Food Chem.*, **17**, 802.
122 Law, J. H., Regnier, F. E., 1971, *Annu. Rev. Biochem.*, **40**, 533.
123 Gordon, H. D., Waterhouse, D. F., Gilby, A. R., 1963, *Nature (London)*, **197**, 818.
124 Pavan, M., 1953, *Arch. Zool. Ital.*, **38**, 157.
125 Waterhouse, D. F., Forss, D. A., Hackmann, R. H., 1961, *J. Insect Physiol.*, **6**, 113.
126 Happ, G. N., Meinwald, J., 1965, *J. Am. Chem. Soc.*, **87**, 2507.
127 Butenandt, A., Hecker, E., 1961, *Angew. Chem.*, **73**, 349.
128 Schneider, D., Kasang, G., Kaissling, K. E., 1968, *Naturwissenschaften*, **55**, 395.
129 Wright, R. H., 1968, *Sci. J.*, **4**, 57.
130 Amoore, J. E., 1965, *Cold Spring Harbor Symp. Quant. Biol.*, **30**, 623.
131 Riddiford, L. M. R., 1970, *J. Insect Physiol.*, **16**, 653.
132 Schneider, D., 1969, *Science*, **163**, 1081.
133 Cmelik, St., 1969, *Hoppe Seyler's Z. Physiol. Chem.*, **350**, 1076.
134 Choudhuri, D. K., Das, K. K., 1969, *Arch. Int. Physiol. Biochem.*, **77**, 609.
135 Schildknecht, H., Weiss, K. H., Vetter, H., 1961, *Z. Naturforsch.*, **17b**, 350.
136 Blum, M. S., Crain, R. C., Chidester, I. B., 1961, *Nature (London)*, **189**, 245.
137 McCullogh, B. T., 1967, *Ann. Entomol. Soc. Am.*, **60**, 862.
138 Dateo, G. P., Roth, L. M., 1967, *Ann. Entomol. Soc. Am.*, **60**, 1025.
139 Blum, M. S., Traynham, I. G., Chidester, J. B., Boggus, I. D., 1960, *Science*, **132**, 1480.
140 Gilby, A. R., Waterhouse, D. F., 1964, *Proc. R. Soc. London Ser. B:*, **162**, 105.
141 Eisner, T., Swithenbank, C., Meinwald, J., 1963, *Ann. Entomol. Soc. Am.*, **56**, 37.

142 Hollande, M. A. Ch., 1909, *Ann. Univ. Grenoble*, **21**, 459.
143 Wain, R. L., 1943, *Annu. Rep. Agric. Hort. Res. Sta. Long Ashton Bristol*, **1943**, 108.
144 Eisner, T., Meinwald, J., 1966, *Science*, **153**, 1341.
145 Roth, L. M., Eisner, T., 1962, *Annu. Rev. Entomol.*, **7**, 107.
146 Park, R. C., Sutherland, M. D., 1962, *Aust. J. Chem.*, **15**, 172.
147 Meijboom, P. W., 1971, *J. Am. Oil Chem. Soc.*, **48**, 684.
148 Hoffmann, G., 1961, *J. Am. Oil Chem. Soc.*, **38**, 1.
149 Meilgaard, M., Elizondo, A., Moya, E., 1970, *Tech. Q. Master Brew. Assoc. Am.*, **7**, 143.
150 Day, E. A., 1965, *The Food Techn.*, **19**, 1585.
151 Farmer, E. H., Sutton, D. A., 1943, *J. Chem. Soc.*, 119.
152 Bolland, J. L., 1949, *Q. Rev. Chem. Soc.*, **3**, 1.
153 Bateman, L., 1954, *Q. Rev. Chem. Soc.*, **8**, 147.
154 Esterbauer, H., Schauenstein, E., 1966, *Fette Seifen Anstrichm.*, **1**, 7.
155 Boyd, E. N., Keeney, E. G., Patton, S., 1965, *J. Food Sci.*, **30**, 845.
156 Jackson, H. W., Morgan, N. E., 1954, *J. Dairy Sci.*, **37**, 1316.
157 Meijboom, P. W., Hoffmann, G., 1968, *J. Am. Oil Chem. Soc.*, **45**, 468.
158 Bergemann, P. H., Koster, J. C., 1964, *Nature (London)*, **202**, 552.
159 Visser, N. K., Lindsay, R. C., 1971, *Tech. Q. Master Brew. Assoc. Am.*, **8**, 123.
160 Forss, D. A., Dunstone, E. A, Ramshaw, E. H., Stark, U. W., 1962, *J. Food Sci.*, **27**, 90.
161 Fleming, H. B., Coob, W. Y., Etchells, J. L., Bell, T. A., 1968, *J. Food Sci.*, **33**, 572.
162 Palamand, S. R., Markl, S., Darvis, D. P., Hardwick, W. A., 1971, *Am. Soc. Brew. Chem. Proc.*, **194**, 230.
163 Forss, D. A., 1969, *J. Dairy Sci.*, **52**, 832.
164 Badings, H. I., 1970, *Proefschrift*, **124**, Nederlands Institut voor Zuivelonderzoek te Ede.
165 Parks, O. W., Keeney, M., Schwartz, D. P., 1963, *J. Dairy Sci.*, **46**, 295.
166 Borovikova, L. A., Roiter, I. M., 1971, *Izv. Vyssh. Uchebn. Zaved. Pishch. Tekhnol.*, **4**, 43 [*Biol. Abstr.*, 13620 (1972)].
167 Pisarnitskii, A. F., Prikladnaya, I., 1965, *Biokhim. Mikrobiol.*, **1**, 144 [*Biol. Abstr.*, 86323 (1966)].
168 Skuniklim, I. M., Nilov, V. V., Ledenkova, T. P., 1965, *Biokhim. Mikrobiol.*, **1**, 675 [*Biol. Abstr.*, 111121 (1966)].
169 Sugisawa, H., 1966, *J. Food Sci.*, **31**, 381.
170 Casey, J. C., Self, R., Swain, T., 1963, *Nature (London)*, **200**, 885.
171 El'Ode, K. E., Dornseifer, D. T. P., Keith, E. S., Powers, J. J., 1966, *J. Food Sci.*, **31**, 351.
172 Swain, T., Hughes, J. C., Linehan, D., Mapson, L. W., Self, R., Tomalin, A. W., 1963, *Proc. Easter Sch. Agric. Sci. Univ. Nottingham*, 160 [*Chem. Abstr.*, **64**, 10324a, 1966].
173 Schormüller, J., Weder, J., 1966, *Z. Lebensm. Unters. Forsch.*, **130**, 158.
174 Self, R., Rolley, H. L. J., Joyce, A. E., 1963, *J. Sci. Food Agric.*, **14**, 8.
175 Mookherjee, B. D., Deck, R. E., Chang, S. S., 1965, *J. Agric. Food Chem.*, **13**, 131.
176 Grey, T. C., Shrimpton, D. H., 1967, *Br. Poult. Sci.*, **8**, 35.
177 Grey, T. C., Shrimpton, D. H., 1967, *Br. Poult. Sci.*, **8**, 23.
178 Kuca, K., Hrdlicka, J., 1966, *Sb. Vys. Sk. Chem. Technol. Praze Oddil Fak. Potravin. Technol.*, **11**, 93 [*Chem. Abstr.*, **67**, 42672j, 1967].
179 Bushkova, L. A., 1966, *Prikl. Biokhim. Mikrobiol.*, **2**, 352.

180 Wyatt, C. J., Day, E. A., 1963, *J. Food Sci.*, **28**, 305.
181 Hughes, R. B., 1963, *J. Sci. Food Agric.*, **14**, 893.
182 Swoboda, P. A. T., Lee, C. H., 1965, *J. Sci. Food Agric.*, **16**, 680.
183 Iwata, N., Morita, M., Ota, S., 1965, *Yukagaku*, **14**, 241 [*Chem. Abstr.*, **63**, 18939g, 1965].
184 Badings, H. T., 1965, *Neth. Milk. Dairy J.*, **19**, 69.
185 Wick, E. L., De Figweiredo, M., Wallace, D. H., 1964, *Cereal Chem.*, **41**, 300.
186 Hrdlicka, J., Janicek, G., 1965, *Sb. Vys. Sk. Chem. Technol. Praze Oddil Fac. Potravin. Technol.*, **8**, 107 [*Chem. Abstr.*, **64**, 18303g, 1966].
187 Hampl, J., Hrdhcha, J., Ocenaskova, A., 1964, *Sb. Vys. Sk. Chem. Technol. Praze Oddil Fac. Potravin. Technol.*, **8**, 131 [*Chem. Abstr.*, **67**, 2183u].
188 Wolford, P. W., Attaway, B. A., Barabas, L. J., 1966, *Citrus Ind.*, **47**, 25, 29, 32, 34.
189 Wickremasinghe, R. L., Swain, T., 1965, *J. Sci. Food Agric.*, **16**, 57.
190 Yamanishi, T., Kobayashi, A., Sato, H., Omura, O., Nakamura, H., 1965, *Agr. Biol. Chem.*, **29**, 1016.
191 Kobayashi, A., Sato, H., Nakamura, H., 1966, *Agric. Biol. Chem.*, **30**, 779.
192 Saijo, R., Kuwabara, Y., 1967, *Agric. Biol. Chem.*, **31**, 389.
193 Drawert, F., Heimann, W., Emberger, R., Tressl, R., 1965, *Justus Liebigs Ann. Chem.*, **694**, 200.
194 Dornandy, D. L., 1969, *Lancet*, **7622**, 684.
195 Goto, T., Kishi, Y., 1968, *Angew. Chem.*, **80**, 417.
196 Eley, M., Cornier, M. I., 1968, *Biochem. Biophys. Res. Commun.*, **32**, 454.
197 Shimomura, O., Johnson, F. H., Kohama, Y., 1972, *Proc. Nat. Acad. Sci. USA*, **69**, 2086.
198 McCapra, F., Hysert, I. W., 1973, *Biochem. Biophys. Res. Commun.*, **52**, 298.
199 Shimomura, O., Johnson, F. H., 1967, *Biochemistry*, **6**, 2293.
200 Bradshaw, J. W. S., Baker, R., Howse, P. E., 1975, *Nature*, **258**, 230.
201 Baker, J. T., Blake, J. D., MacLeod, J. K., Ironside, D. A., Johnson, I. C., 1972, *Aust. J. Chem.*, **25**, 393.
202 Kornberg, H. L., Elsden, S. R., 1961, *Adv. Enzymol.*, **23**, 401.

Index

Acrolein
  inhibition of biological oxidation by 2
  and protein cross-linking 32
  and toxicity 33, 35
'Aerial phytoncide' 40
Affinity of hydroxyenals for sulphydryl groups 75
Alarm pheromones 188
Alcohol intoxication, and aldehydes 174
Aldehyde derivatives, biological action 158
Aldehyde hydrogen sulphite amine complexes in tumour therapy 128
Aldehyde hydrogen sulphite complexes in tumour therapy 128
Aldehydes
  from autoxidation of unsaturated fatty acids 192
  biochemical interrelationships 166
  in bioluminescence 194
  and cuticular leaf growth 181
  degree of hydration 9
  as flavour components in food 189, 193
  mechanism of action by 4
  occurrence *in vivo* 3, 181, 184
  and odour 184
  and plant waxes 181
  and protein biosynthesis 11, 18
  reactive sites of 3
  role in schizophrenia 174
  as semiochemicals 187
Aliphatic aldehydes, inhibitory effect on protein biosynthesis 11
Alkanals
  antitumour effects of 17
  and bacteria 20
  and fungi 20
  physiological effects of 15
  reactions with amino groups 4, 9
  reactions with proteins 11
  reactions with sulphydryl groups 4
  structure in water 9
2-Alkenals
  reactions with amino groups 5, 31
  reaction involving cysteine 6
  reaction with glutathione 6
  reactions with sulphydryl groups 4, 25
  reactivity of different 28
  structure 25
Allyl formate, and liver damage 35
β-Aminopropionaldehyde, conversion to MA 140
Anticancer agents 52

Antileukaemia action of thiosemicarbazones 159
Antimicrobial action of thiosemicarbazones 40
Antitumour activity
  of aldehyde derivatives 159
  of alkanals 17
  of alkenals 35
  and chemical constitution 18
  of formaldehyde 20
  of glyceraldehyde 1, 107
Antiviral activity
  of aldehyde derivatives 158
  of enediols 132
  of α-hydroxyaldehydes 132
  of α-ketoalcohols 132
Antiviral effects
  of di- and triketones 132
  of α-ketoaldehydes 118, 131
  of α,β-unsaturated aldehydes 40
Aroma of fruits 184
Autoxidation of linoleic acid 190
Autoxidized polyene fatty acids 42

Bactericidal action
  of citral 39
  of α,β-unsaturated aldehydes 38
Bacteriostatic effect, α-ketoaldehydes 1
Biogenic α-ketoaldehydes, as naturally occurring growth inhibitors 2
Bioluminescence and aldehydes 194
Biosynthesis of DNA, and D,L-glyceraldehyde 104, 106
Biosynthesis of RNA, and D,L-glyceraldehyde 104, 106
Blockage of arginine residues in proteins by α-ketoaldehydes 115

Cancer cell growth 128
Carbonyl compounds in fruits 185
Carcinostatic action of thiosemicarbazones 159
Cell division, inhibition by formaldehyde 21
Cell respiration
  effect of alkanals on 16
  and HPE 71
Ceroid, age pigment 136
Chemical disinfection, and glutaraldehyde 146
Chemotherapy
  of cancer 108
  of growing tumours 127

Citral
    antitumour effects of 2
    and bactericidal activity 39
    natural occurrence 184
Collagen
    cross-linking of 14, 117
    and manufacture of leather 13
Collagenaldehyde 181
Cross-linking of proteins 14, 32, 117, 140, 143
Crotonaldehyde, and plant viruses 41
Cyanopsin 179
Cyclophosphamide
    as agent in tumour therapy 2, 37
    degradation 37
Cysteine, reaction with 2-alkenals 6
Cytostatic agents 57

Dehydroretinal (vitamin A_2) 179
3-Deoxyglucosulose
    inhibition of cell growth 168
    from liver 168
Diacetyl, bacteriostatic effect 130
1,4-Dihydroxypentene, biogenesis from HPE 96
α-Diketones, and amines 113
Diphtheria toxin, deactivation by glyoxal 1
DNA, reversible interaction with glyoxal 120
Dodecenal in bioluminescence 194

Ehrlich ascites tumour cells
    and glycolysis 43
    and respiration 43
    and sensitization by HOE 52
Energy metabolism, and HPE 55
Enzymes
    inhibited by glyoxal 121
    inhibited by methylglyoxal 120
Epithelial tumours, and glyceraldehyde 105
β-Ethoxy-α-oxobutyraldehyde
    (see Kethoxal)

Fatty aldehydes from animal tissues 163
Formaldehyde
    action on proteins 11
    and cell division 21
    and cell respiration 16
    and lipase 14
    and serum albumin 14
Fungicidal action, and chemical reactivity 39

Fungicidal effects, of α,β-unsaturated aldehydes 38
Fusel oils 17

Gelatine, cross-linked with glyoxal 117
Gliadin, cross-linked with glyoxal 117
Glucose metabolism, and
    D,L-glyceraldehyde 105
Glutaraldehyde
    and chemical disinfection 146
    cross-linking effect on proteins 140
    and deactivation of viruses 146
    general features 140
    in leather manufacture 140
    reactions with peptides 142
    reactions with proteins 142
    structure in water 141
Glutathione
    and diethyl maleate 88
    and methylglyoxal 115
    model reaction with HOE 48
    and α,β-unsaturated carbonyl compounds 89
Glutathione-S-alkene transferases 89
Glyceraldehyde
    activity spectrum 109
    and bacteria 106
    conversion to methylglyoxal 107
    inhibition of glycolysis by 1, 103, 104
    and tumours 1, 105, 107
    and viruses 106
D-Glyceraldehyde, cytostatic action of 108
D,L-Glyceraldehyde
    and DNA biosynthesis 104, 106
    and glucose metabolism 105
    and protein biosynthesis 103
    and RNA biosynthesis 104, 106
Glycolysis, and L-glyceraldehyde 1, 103
Glyoxal
    and arginine residues 115
    cross-linking with proteins 117
    deactivation of diphtheria toxin by 1
    inhibition of cell growth 123
    inhibition of DNA synthesis 123
    inhibition of enzymes 121
    inhibition of glycolyses 121
    inhibition of oxygen uptake 121
    inhibition of protein syntehsis 122, 123
    inhibition of RNA synthesis 123
    in irradiated foodstuffs 164
    and nucleotide sequence of tRNA 119
    in plant products 164
    reversible interaction with DNA 120

Glyoxal (continued)
   specific for guanine bases 118
   structure in water 112
   thiosemicarbazone as bacteriostatic agent 131
Growth regulators 121
Guanylhydrazones of α-ketoaldehydes, carcinostatic effects 159

Haemoglobin, glyoxal-treated 116
Heptanal, and tumours 17
2-Hexenal, and 'aerial phytoncide' 40
2-trans-Hexenal
   from ethyl linolenate 191
   as fungicide 184
   natural occurrence 183
Homologous hydroxyenals, comparison of activities 53
Hydration of aldehydes 9
α-Hydroxyaldehydes 103, 132
4-Hydroxy-2-alkenals
   and biochemical or biological effects 52, 66
   inhibitory effects 2, 49
   occurrence 42
   synthesis 51
4-Hydroxyenals from 2-deoxyribose 87
5-Hydroxyindol-3-ylacetaldehyde, formation from tryptamine 172
4-Hydroxy-trans-2,3-octen-1-al (HOE) 46
   and glycolysis 57
   inhibitory effects of 46, 49
   model reaction with glutathione 48
   reaction with sulphydryl groups 49
   as selective anticancer agent 52
   and sensitization of EATC 52
4-Hydroxy-2-oxobutanal as inhibitor 125
4-Hydroxy-2-oxobutyraldehyde, inhibition of bacterial growth 169
4-Hydroxypentenal (HPE)
   binding to proteins 61
   and biosyntheses of nucleic acids 71
   and cell respiration 71
   conversion in vitro to 1,4-dihydroxypentene 96
   and energy metabolism 55
   and glycolysis 57
   and inhibition of respiration 58, 72
   and losses of sulphydryl groups in proteins 66
   and macromolecular biosynthesis 55
   metabolic fate in vitro 95
   metabolic fate in vivo 94

4-Hydroxypentenal (continued)
   and mitosis 75
   and nucleic acid biosynthesis 59
   and protein biosynthesis 71
   and smooth muscle (uterus) 85
   and striated muscle (diaphragm) 85
   and succinate oxidation 74
   and sulphydryl groups 60
   and thymidine incorporation into cells 82
   and tumour metabolism 57
   and tumour therapy 77
4-Hydroxypentenal-cysteine
   (1:1) adduct 81
   (1:2) adduct 82
4-Hydroxypentenal diethyl acetal, metabolic fate of 91
Hyperthermia 53, 110

Incorporation of thymidine into DNA 82, 103
Incorporation of uracil into RNA 103
Indol-3-ylacetaldehyde, formation from serotonin 172
Influenza virus, and α-dicarbonyl compounds 106
Inhibition
   of bacterial growth by 4-hydroxy-2-oxobutyraldehyde 169
   of cell division by formaldehyde 21
   of cell growth by 3-deoxyglucosulose 168
   of cell respiration by HPE 58, 71
   of glycolyses by glyoxal 121
   of nucleic acid biosyntheses 59, 71
   of protein biosyntheses 71
   of succinate oxidation 72
Inhibitory effects on energy metabolism, by 4-hydroxy-2-alkenals 2
Inhibitory substances and lipid hydroperoxides 45
Insulin, arginine blocking in 116
Iodopsin 179
Irradiated foodstuffs, and biological activity 42
Isatin-3-thiosemicarbazone, antiviral action 158

Kethoxal (β-ethoxy-α-oxobutyraldehyde)
   and amino acids 113
   inhibition of protein synthesis 122
   and nucleotide sequence of tRNA 119
   specific for guanine bases 118

α-Ketoaldehyde dehydrogenase 129
α-Ketoaldehydes 112, 115
  and amines 113
  and amino acids 113
  as antileukaemic agents 127
  antitumour effects of 2, 128
  antiviral properties of 2, 131
  and bacterial growth 1, 130
  inhibition of protein synthesis 123
  metabolic fate 129
  and nucleic acids 118
  and proteins 115
  as regulators and inhibitors of cell growth 1, 121
  and tumour growth 126
  and uptake of oxygen 120

Leather, and glutaraldehyde 140
Length of aliphatic chain, and activity of hydroxyenals 53
Light-sensitive pigments of the retina 179
Linoleic acid, autoxidation 190
Lipase, and formaldehyde 14
Lipid hydroperoxides 43
Lipofuscin, age pigment 136
Losses of sulphydryl groups in proteins 66
Luciferin-luciferase system 194
Lysozyme, arginine blocking in 116

Macromolecular biosynthesis, and HPE 55
Malonaldehyde (MA) 133
  and ageing of cells and organs 139
  from β-aminopropionaldehyde 140
  and autoxidation of lipids 133
  and cysteine 136
  essential for cell growth control 140
  and liver damage 140
  metabolic fate 133
  and methionine 136
  and oxygen uptake 139
  reactions with amino acids 133, 135
  reactions with nucleic acids 133, 136
  reactions with proteins and enzymes 133, 137
  tautomerism with β-hydroxyacrolein 134
Methylglyoxal
  in animal tissues 1, 164
  enzymatic formation in tissues 165
  in flavouring matter 164
  and glutathione 115
  from glyceraldehyde 107
  inhibition of enzymes 120

Methylglyoxal
  nonenzymatic formation from dihydroxyacetone 165
  nonenzymatic formation from glyceraldehyde 165
  as normal metabolite of *E. coli* 130
  structure in water 112
1-Methylisatin-3-thiosemicarbazone, antiviral action 158
Mitochondria, stabilization of their configurational state by glutaraldehyde 145

Natulan (cytostatic agent) 19
2-*trans*,6-*cis*-Nonadienal, from ethyl linolenate 191
Nonprotein thiol compounds
  and HPE 66
  protective function of 69

Occurrence of aldehydes 163
Oxidoreductase and α,β-unsaturated ketones 90

Phenylglyoxal, and arginine residues 115
Pheromone recognition 186
Pheromones 185
Plant viruses, and crotonaldehyde 41
Plasmologens 163
Porphyropsin 179
Promine, cell growth accelerating factor 122, 125
Protection of sulphydryl groups of soluble proteins 69, 74
Protein biosynthesis, inhibitory effects of aldehydes on 11, 18, 104, 123
Protein-bound aldehydes, determination of 13
Protein conjugates, and glutaraldehyde 144
Proteins
  effect of alkanals on 11
  effect of HPE on 61
Pyridine-2-carboxaldehyde
  thiosemicarbazone, antileukaemia effect 159
Pyridoxal
  biosynthesis 175
  importance of 174
  from pyridoxol (vitamin B_6) 174
Pyridoxal phosphate
  as coenzyme 174
  mechanism of catalysis 175

Index

Reagents for specific modification of arginine residues in proteins 115
Recruiting pheromones 188
Reduction of α,β-unsaturated ketones 90
Retinal (vitamin A_1) 179
  biosynthesis 180
'Retine', cell growth inhibitory factor 122, 125
'Retine'-active biogenic α-ketoaldehydes 124
Rhodopsin 179
Ribonuclease, and MA 138

Schizophrenia, role of aldehydes 174
Semiochemicals 185
Serotonin
  metabolism 15, 172
  physiological effects 172
  and sleep 172
Serum albumin, and formaldehyde 14
Sex pheromones 188
Site of attack in substrates, by aldehydes 3
Sleep 172
Specificity of hydroxyenals 87
Spermidine aldehyde, antiviral activity 40
Spermine aldehyde, antiviral activity 40
Stalagmoptysis 51
Stalagmosis 51
Succinaldehydic acid
  metabolic pathways 172
  physiological action 172
Succinate oxidation, inhibition by HPE 72
Sulphydryl enzymes, and inhibitions caused by HOE 49
Sulphydryl groups and hydroxyenals 75
Sulphydryl losses in proteins 66
Synergistic action of aldehydes with hyperthermia 53, 110

Thiosemicarbazones
  and antimicrobial action 40, 131
  and bacteriostatic properties 158
  and carcinostatic properties 159
  and tuberculostatic activity 158
Thymidine incorporation, and HPE 82
Thymidine incorporation into DNA, and glyceraldehyde 103
Tryptamine metabolism 172
Tumour inhibition (see Antitumour activity)
Tumour therapy
  and cyclophosphamide 2
  and hydroxyenals 77

Universal mechanism regulating growth 121
α,β-Unsaturated aldehydes
  antiviral effects of 40
  bactericidal effects of 38
  fungicidal effects of 38
  metabolic fate of 88
  structure of 25
  toxicity of 2, 32
  and tumours 36
α,β-Unsaturated carbonyl compounds
  cytotoxic effects 150
  and glutathione 89
α,β-Unsaturated ketones, reduction 90
Uracil incorporation into RNA, and glyceraldehyde 103

Visual pigments 180
Vitamin A, in growth, development, and reproduction 180
Vitamin A_1 (retinal) 179
Vitamin A_2 (dehydroretinal) 179
Vitamin B_6 (pyridoxol) 174